Social and Cultural Studies of Robots and AI

This is a groundbreaking series that investigates the ways in which the "robot revolution" is shifting our understanding of what it means to be human. With robots filling a variety of roles in society—from soldiers to loving companions—we can see that the second machine age is already here. This raises questions about the future of labor, war, our environment, and even human-to human-relationships.

More information about this series at
http://www.palgrave.com/gp/series/15887

Kathleen Richardson

Challenging Sociality

An Anthropology of Robots, Autism, and Attachment

palgrave
macmillan

Kathleen Richardson
Faculty of Technology
De Montfort University
Leicester, UK

ISSN 2523-8523 ISSN 2523-8531 (electronic)
Social and Cultural Studies of Robots and AI
ISBN 978-3-030-09069-2 ISBN 978-3-319-74754-5 (eBook)
https://doi.org/10.1007/978-3-319-74754-5

Printed on acid-free paper

This Palgrave Macmillan imprint is published by the registered company Springer International Publishing AG part of Springer Nature.
The registered company address is: Gewerbestrasse 11, 6330 Cham, Switzerland

For my witness and the little ones

Foreword

> The attitude of man is twofold, in accordance with the twofold nature of the primary words which he speaks.
>
> The primary words are not isolated words, but combined words.
>
> The one primary word is the combination I-Thou.
>
> The other primary word is the combination I-It; wherein, without a change in the primary word, one of the words He and She can replace It. (Martin Buber, I and Thou, 1937)

The ideas for this book were first conceived between 2003 and 2004 when I carried out my doctoral fieldwork in social anthropology at the labs at the Massachusetts Institute of Technology. It was here that I first came across analogies being made between robots and autism. I noticed that studies of disability, illness and difference were drawn on by these robotics researchers when they were making their humanoid robots. These researchers began to explore whether autism might give roboticists an insight into the 'mind' of a robot. One such paper called 'Implementing Models of Autism with a Humanoid Robot' took concepts from autism into robotics and back again. The growing popularity of the 'social' robot (a robot whose primary function is to be socially interactive), meant that roboticists began to ask questions about what is and what is not social. I was interested in how ideas of what constitutes sociality are personally and culturally informed. Are we really able to agree that a machine is 'social' if it has a face or can perform some interactive behaviour? My work into autism and robots first began here because it was in these labs that I was first confronted with the idea that a machine could be a social companion, a friend, or a therapist. It was a few years later that I was exposed to the idea that robots could become girlfriends or wives. Robots performing intimate or relational roles has long been a staple of fiction—since long before a technological programme of action was initiated.

In my own studies I found robotics scientists would draw on the fields of disability and difference to use as a reference point to make robots, as if people with disabilities and differences are somehow not fully human and can be used as a model of comparison.

There are many popular theories that ignore species-specific sociality approaches to making sense of our world and propose we are cyborgs or multi-species. They make analogies between humans and machines or downgrade human interpersonal relationships and suggest important human relationships (a companion, lover, friend or therapist) can be replaced by robotic machines. Some propose that it is because human intimate relationships are so degraded today that people are becoming unable to differentiate between a human and a robot. Some are arguing the opposite, that robotic machines are becoming so sophisticated that it is becoming impossible to tell the difference between a robotic machine and a human.

Robots and AI are betwixt and between because they are believed to somehow cross a threshold between the world of humans and human-made artefacts. Robots and AI are thought to be something 'more than' ordinary artefacts: they can think, feel and move. This idea is more compelling if the robot looks humanlike.

Children and adults with autism are also believed, by some, to be betwixt and between, between the social and the asocial—a sphere that is written about by psychologists such as Simon Baron-Cohen. There has been a move to join these concepts and new partnerships have developed between the biomedical psychology of autism and the therapeutic benefits of robots and AI.

My central argument in this book is that all within and between human beings flows into the human-made. The human-made (i.e., robots), do not shape people; they reflect them. However, if what we produce in our artefacts reflects back to us an empty egocentric illusion informed by lack of empathy and anti-humanism, this is a problem. Who controls the making and production of the artefacts has more control over the environment we share and the relations that people can have between each other.

Real change comes from those interpersonal relations within and between people; if we can change those underlying relations, then the artefacts produced will change. The power is within and between I and you and in our human bonds and ties.

Leicester, UK
3 November 2017

Kathleen Richardson

Preface

> The Thou meets me through grace – it is not found by seeking.
>
> The Thou meets me. But I step into direct relation with it. Hence the relation means being chosen and choosing, suffering and action in one; just as any action of the whole being, which means the suspension of all partial actions and consequently of all sensations of actions grounded only in their particular limitation, is bound to resemble suffering.
>
> The primary word I-Thou can only be spoken with the whole being.
>
> All real living is meeting. (Martin Buber, I and Thou, 1937)

The term 'social' is a complex one, with multiple meanings. In social anthropology, the term 'social' refers to the 'social order'—the structure which shapes people's lives. In anthropology in the United States (U.S), 'the cultural' is often used instead of 'the social' to mean almost the same thing (though many an anthropologist would argue about the intricate differences). However, there is an important area of note, which is perhaps why we talk about 'social robots' rather than 'cultural robots' despite the fact that most of the robots I studied were built in the U.S. Often the term 'social' is implicitly synonymous with *socialising*, something people do as a shared experience involving enjoyment and entertainment. The term 'social' can also mean an interpersonal interaction. The social can refer to the macro—the structure of 'society' or shared experience —or the micro —interpersonal relationships. The social traditionally was about human beings and what they did with each other. But this has changed to now incorporate human-made artefacts such as robots.

In this book, I draw on the meaning of the social as an interpersonal phenomenon, a meeting between myself and different kinds of people. Some of this research is drawn from meetings with children with autism or autistic children (depending on your preferred terminology, something *you have*, or something *you are*).

In autism literature, the role of the interpersonal relationship is diminished, so much so that it is now commonplace to suggest that children with autism prefer things (including robots) to other people. There is a community of researchers that now claim that children with autism can respond well to robots with social features (outwardly appearing like humans) because people are too complex. For these robotics researchers, a humanlike robot is a 'simplified person'. How did we get to the idea that robots could be simplified persons? Or that children with autism prefer humanlike robotic machines over people? This is the subject of this book.

There is always reason for caution when writing about a subject matter that touches on matters of disability and difference. What authority does a person without autism have to write about autism? There can be no easy answers here. I became curious about autism because of the ways in which robotics scientists use autism to make sense of what it might be like to be a robot. It was exploration of this idea that led me to carry out this research. I welcome further discussion and dialogue with adults with autism, parents, activists, social scientists and roboticists on the issues I discuss in this book.

The widespread use of the empathising-systemising theory of essential sex differences in the development of robots for children with autism is also something I want to challenge in this book. I want to explore the way that autism has become a way to talk about typical masculinity, as if autism were an extreme form of masculinity that explains away male violence and lack of empathy. In my experience, people with autism do not lack empathy, and they are not deficient in empathy in the way of males who are socialised to hide their feelings, use violence as a way to control others, or act out violence on others. Consequently, I will be exploring analogies made between men, machines/robots and autism.

Autism researchers note two interconnected behaviours in children with autism: echolalia and pronoun reversal. 'Echolalia' comes from the Roman myth—Echo only repeats the last words spoken to her; she is tormented by an inability to respond in her own way. Pronoun reversal refers to the speaker (in the 'I' position) using the pronoun of the listener (in the 'You' position). So the child will say 'you sat down' to refer to the subject of the action (her/him) taking the action (sitting down). However, the listener will still be holding the you position. A you meets a you, in effect. What does it mean when a you meets a you? And if the child is speaking about himself from the pronoun position of you, what has happened to his or her I?

This difficulty in expressing regular pronouns is often dismissed as a language error (the child knows which pronoun is correct but, on occasion, makes mistakes), or irrelevant behaviour (one of many language errors that are part of the repertoire of communicative speech). But what if the pronoun reversal actually tells us something about the experience of the child—that the child experiences themselves not as a separate person in the position of the I-speaker, but as part of the you-listener? Perhaps in this linguistic moment an insight into the ontological experience of the child is revealed.

If we took this into account in sense-making of autism, perhaps we would better understand the way in which children with autism need to regulate social interaction when confronted with a concern about their own selfhood and the boundaries between themselves and others. Our interpersonal encounters help us to make sense of ourselves, and if this is impaired, then it impacts on selfhood and ways of knowing others and being known by others. It should come as no surprise that children and adults with autism suffer from extreme anxiety and panic. In attachment studies one role of caregivers is to assist a developing child to regulate their emotions by soothing and comforting the child when he or she is in distress. The child internalises this soothing and comfort and can make use of it when the parent is not there. This is a way in which intimate interpersonal attachment is crucial for human development—especially so for children.

I would like to thank the British Academy for their funding support for this project and all those people who have helped contribute to the research by sharing their time and resources. I would like to thank the European Union for their funding support and helping me to learn more about how the research is developing and allowing me to at least contribute to the ethics of this area. For the last three years I have been working as an ethics advisor on DREAM (Development of Robot-Enhanced Therapy for Children with Autism Spectrum Disorders). DREAM is dedicated to developing robot technologies for helping psychotherapists in the field of Applied Behavioural Analysis (ABA) approaches to autism. I try as much as possible in my role as ethics advisor to introduce autism advocacy and critical autism perspectives into my ethical evaluations.

I would like to thank the school in London and all the staff and students I met and spoke with. You work so tremendously hard, and the staff struggle to get resources for the students. I would also like to thank the robotics scientists and psychologists. In keeping with issues of anthropological commitments to confidentiality, I have the people, the places and the robots pseudonyms. As the robotic scientists and psychologists I have worked with produce artefacts such as scholarly books, papers, dissertations and robots, I have only referred to public material if such activities do not conflict with my initial commitment to honour the relationships of my interlocuters, many of who I continue to meet as colleagues. All the interlocuters showed great genorosity by sharing their work with me. However, I do take responsibility for my critical reflections on some of these processes in the field of robots and autism.

I would like to thank the editors at Palgrave Macmillan for supporting this project.

I draw heavily on the research conducted by the Critical Autism Studies network and their brave work in questioning normative power-structures. The Critical Autism Studies network is inviting us to look beyond labels, or brains, or systems.

This book has been a journey from being mostly within mechanical notions of the human to rejecting these approaches in their entirety. There is no machine in the human being. Reproducing this fiction through making analogies between children or adults who are labelled autistic and robotic machines is a serious problem, not least because it deprives human beings of species-specific sociality and humanity.

I have some people to thank directly. Firstly, my very special thanks to Florence Gildea, who helped to prepare the manuscript and was a much-needed ally at the end of the process.

My understanding of I-you interrelatedness is drawn from my many dialogues with psychodramatist Caitlin Buon. She is extraordinarily intelligent and gifted and has an unyielding kindness that is as powerful as any great army!

Most of all, I would like to thank the young students of the school who put up with a stranger coming in and offering toys, a robot and a crane, to play with. I am sorry if I annoyed you in any way and my understanding of what you might be experiencing was not as well developed as it could have been. I still have a lot to learn and you have been my teachers. I am so grateful to all of you.

Leicester, UK — Kathleen Richardson

Contents

CHAPTER 1

Introduction: Challenging Sociality

1.1 Autism and the Machine

Social interaction and communication are central features of human sociality, but could robotic machines be social? And what of those diagnosed with autism? What kinds of lifeworlds emerge for those who struggle to grasp the complexity of social interaction? What does sociality mean for those humans who find social interaction, so necessary for survival and happiness, confusing and difficult? This book explores the role of intermediaries designed to therapeutically help children with autism develop social skills. In the early twenty first century we are on the cusp of a new technologically mediated sociality, whereby robots (anthropomorphic machines) are produced to substitute humans in interpersonal relationships. As anthropomorphic robots take on these roles, particularly in fields such as healthcare, robots are increasingly being used as therapeutic tools for helping adults and children with mental health, physical or social conditions. These 'social' robots, humanlike in appearance and behaviour, are used as therapeutic tools for children with autism, and this reveals something about human-robotic imaginings.

Human relationships are characterised by a developmental awareness of human sociality—that is an ability to make sense of, and form, reciprocal social attachments with others. Children with autism, at least according to the psychological science literature, display difficulties in developing social awareness of others and experience difficulty in forming attachments, even to their primary caregivers (with whom human beings typically develop

K. Richardson, *Challenging Sociality*, Social and Cultural Studies of Robots and AI, https://doi.org/10.1007/978-3-319-74754-5_1

their core early relationships). Children with autism have difficulty reading the 'social' cues of others. Thus they may struggle to maintain eye-contact, read facial expressions and understand their own feelings. Children with autism are also said to prefer objects to other persons. When Leo Kanner first described autism in his 1943 paper, he noticed the interest in the physical world displayed by these children. The children's desire for 'aloneness' was connected to their interest in objects, he suggested.

> The outstanding, 'pathognomonic,' fundamental disorder is the children's inability to relate to themselves in the ordinary way to people and situations from the beginning of life. (Kanner 1943, p. 242)

Kanner continued along these lines, conflating the particular relationship the children expressed to objects with their 'aloneness':

> Objects that do not change their appearance and position, that retain their sameness and never threaten to interfere with the child's aloneness, are readily accepted by the autistic child. He has a good relation to objects. (p. 246)

Though research in autism studies has developed significantly since the 1940s, the two main features of Kanner's argument—'an inability to relate' and a 'good relation to objects'—underlie the arguments for using anthropomorphic robots to help children with autism develop social behaviours.

In what follows, I explore how machines that take on anthropomorphic forms in the guise of robots are used to help children with autism develop sociality. To understand how robots can take on these roles for children with autism, it is necessary to understand both the psychiatric context in which autism is understood as a specific kind of condition, and the field of social robotics. Though the field of social robotics is shaped differently from this psychiatric context, the two fields have formed a partnership around the topic of autism.

Autism and social robots encourage us to rethink the meaning of the social in contemporary anthropological theorising and in robotics. What model of the 'social' is employed in these narratives? What kind of 'relationship' is invoked through these meanings? The intermediaries that I study are primarily humanoid robots, but also toys, computer programmes and cartoon animations.

I will explore how intermediaries are developed to bridge this divide between the social and asocial fields. But I suggest that the mechanical models of the social that underscore Euro-American biomedical psychiatry must be understood before the argument that machines can help children with autism is accepted.

The data that informs my project comes from two years of participant observation and interviews with autism experts (including psychiatrists, educationalists and roboticists), and my participation in a unique psychological experiment exploring the effectiveness of a humanoid robot for helping children with autism develop social behaviours. The experiment was conducted at a specialist school for children with autism. I have also spent additional years observing how the field of autism and robots has grown and changed through my role as an ethics advisor on a European project called DREAM (Development of Robot-Enhanced Therapy for Children with Autism Spectrum Disorders).

The ethnographic and experimental data that is written about in this book is from research material collected during my British Academy Postdoctoral Fellowship (2009–2012), held at University College London. As part of this research I participated in an experiment testing the effectiveness of a robot for children with autism. The experiment was conducted over three months in 2011. The experiment was designed by an autism expert, and a robot for the experiment was provided by the robotics group at a British university. The experiment involved running a play protocol and then encouraging playful interactions between children with autism and a robot or a toy crane. The video-taped data was then coded by an independent researcher. When the data was analysed, there was no significant difference in social behaviours among children who interacted with the robot or those children who interacted with the crane. The group put this down to a failing in the 1⊠1 experiment, it was not rigorous enough. The results of the experiment were considered void (from the 'scientific' perspective) and never published.

I examine this experiment and the experimental process here as, unlike scientists, anthropologists can use the data from 'failed' experiments. Failed experiments tell us about the way in which science is constructed and what 'truths' it allows. Where do all those failed experiments go? What does it mean to have a worldview that is composed not really of what you find, but what you decided is experimentally valid? To situate my fieldwork within anthropological studies of sociality, I use these studies to reflect on other kinds of knowing about truth.

In line with arguments proposed by Critical Autism Studies researchers, who study the biopsychosocial contexts in which knowledge is produced about people with autism, I propose that the view that robots are useful for aiding children with autism can only make sense within a machine model of autism. This model flattens out distinctions between humans and machines.

My position goes against the current thinking in social anthropology that promotes 'flat-ontologies' and cyborgs, according to which, humans and nonhumans are read as equivalent, like in a mathematical formula. Mathematical formulas are important here and show us how a principle of equivalence between humans and machines occurs in the current framing of therapeutic robots for children with autism. The social robot is the intermediary between human and machine. Intermediaries act as transporters, moving meanings from one place to another, but such intermediaries can only exist if there is an underscoring principle of equivalence to equate them together.

The topic of autism and robots holds an attraction for contemporary anthropologists interested in sociality. Sociality is that field of life focused on our interconnections with others (direct relations and machine-mediated human relations) and how we make sense of those interconnections. Life today is marked by an interest in sociality, from the dissolution of boundaries between humans, animals and artefacts (Haraway 2003; Latour 2005) to technological engagements with sociality epitomized by social networking and social robots. The social, then, is a defining concept of the contemporary age. What sociality is, and to whom it belongs, is a controversial question. To some, a dichotomy between social-humans and nonsocial-machines makes less sense in today's digitally connected world, where humans and machines become integrated communication systems, as the 'data' packets (text, audio, image, video) are exchanged between people via the internet and new machine mediated communication systems. To what extent is the growth of digital culture about a depersonalised sociality? To what extent might digital culture emerge out of the autistic 'extreme male brain'? Others write about the corporate ideology that underscores digital culture that privileges instrumental neoliberal values over affective ties (Boellstorff 2015; Dourish and Bell 2011; Gershon 2010; Helmreich 1998; Kelty 2005; Pearce et al. 2011). Pioneers of digital culture, such as Mark Zuckerberg and Bill Gates (along with many other luminaries of the digital age) are thought to be on the autism spectrum.

In the Silicon Valley area of California some of the highest incidences of autism spectrum diagnoses are reported. Autism advocate and campaigner Temple Grandin suggested the same when she wrote:

Steve Jobs was probably mildly on the autistic spectrum. Basically, you've probably known people who were geeky and socially awkward but very smart. When does geeks and nerds became autism? That's a gray area. Half the people in Silicon Valley probably have autism.[1]

The figures for prevalence of autism differ widely though, as Timimi and McCabe point out, depending on which measurement is used. Some figures put the incidence of autism at 1 in 50 children, 'a 50-fold rise in just four decades' (2016a, p. 30). What has led to this huge increase in diagnoses of autism? Is there more autism today? Or better diagnostic techniques? Has a culture that has increased our dependency on machines for everyday actions reduced our level of interpersonal social contact? Is it related to diet? Is it a result of modern pharmaceuticals, pesticides or vaccines? Or is parenting responsible?

This book charts the rise and use of robots for helping children with autism develop social skills. Autism is the name given by biomedical and psychological clinicians to the condition affecting a category of persons who may show marked impairments in social-communication and interaction (Grinker 2008). This collection of impairments includes: a disinterest in primary and secondary caregiver attachment figures; a failure to recognise people or bond 'appropriately' to a caregiver (i.e. engage in reciprocal communication and interaction) (Baron-Cohen and Wheelwright 2004, pp. 163–175; Baron-Cohen 2011); a preference for certain kinds of things (Baron-Cohen 2004; Baron-Cohen et al. 2009c); and high levels of distress and anxiety (Kim et al. 2000, pp. 117–132). Autism is often described as a 'spectrum' and understood by experts in the field as a heterogeneous category involving a variety of persons with social interaction difficulties which can take many forms—from an uncertainty in labelling emotions to lack of eye-contact (Baron-Cohen et al. 2009a, pp. 3567–3574; Castelli 2005, pp. 428–449). As Timimi and McCabe point out, this is picking and choosing who has and who does not have autism according to the model of the practitioner rather than empirical evidence. But perhaps this represents an opportunity. The lack of any actual evidence for a neurological, biological basis of autism should encourage researchers to think again about what should be done in relation to 'autism' and whether it even exists at all as a discrete identifiable category. Timimi and McCabe (2016a, b) have this to say:

[1]Temple Grandin Website [Online] Accessed 19.8.17 <http://www.templegrandin.com/>

> Not only do the 'core' features such as difficulties in social communication have a large cross-over with people who would not consider themselves as having a 'disorder', but autism as a mental state (referring back to Bleuler's original use of the term autism to denote a state of social withdrawal) and elements of autism symptoms as traits are common in a verity of mental health diagnoses from ADHD to depression and anxiety to psychosis. No wonder it is difficult to design a specifically 'autistic' intervention for the diagnoses (p. 39).

Indeed, anthropological research is critical of these biomedical assumptions regarding the nature of autism as a 'fact' or 'thing' that exists across time and space, pointing to different cultural assumptions held about autism, its diagnosis (Grinker 2008, 2010) and proposed causes (Kaufman 2010). Moreover, pioneering work in this area is using autism as a means to explore anthropological meanings of sociality, with the research of Ochs and Solomon (Ochs and Solomon 2010; Ochs et al. 2004) pointing to an 'autistic sociality' suggesting a broader approach to understanding sociality than that developed by psychologists and the biomedical community.

Critical Autism Studies (CAS) has emerged as a new field, pioneered by a collection of researchers and activists who are critical of biomedical models of autism, the commodification of autism, and the label of autism and its impact on the lived lives of human beings with differences (Runswick-Cole et al. 2016; Timimi and McCabe 2016a, b; Runswick-Cole 2016; Russell 2016). CAS presents a challenge to the scripts of biomedical approaches to autism and offers new ways of making sense of being human in different ways. In *Re-Thinking Autism: Diagnosis, Identity and Equality* (2016), which outlines the goals and research agenda of CAS, Runswick-Cole, Mallett and Timimi write:

> The recurrent failure of any visible progress in academic attempts at discovering what autism 'is' serve to obscure a simpler and more likely possibility – that the reason science is not uncovering what autism is, is because it doesn't exist at a biological level. If, as seems likely, this is the case, then autism can only be understood through examining 'it' as a socially/culturally produced phenomenon. (2016, p. 9)

The problem though, as Critical Autism Studies (CAS) researchers know well, is that to critique 'autism' is to be inside a particular kind of autism narrative—the biomedical one that dominates understanding, and the distribution of resources and service provision. Katherine Runswick-Cole (2016) is an academic and a mother of a child with autism. As a parent and an academic she reflects on how parents participate in autism narratives that span the 'gifted' to the severely disabled child. Runswick-Cole explores if there are other ways of knowing her son, and others with an autism diagnosis that are outside the biomedical paradigm, and how alternatives of selfhood and can be produced she writes 'my journey with autism pre-dates the birth of my son…Mothers of disabled children find themselves in a troubling space; they have often been vilified by the disabled people's movement as oppressors of their disabled children while simultaneously being pathologised by the professions as "grief stricken" or "in denial" in "coming to terms with" having a child who is marked as deviant or disordered' (p. 20). CAS researchers attempt to reframe the discussion of what it means to be autistic, and be in the world, when the dominant paradigm is driven by a powerful and well-funded biomedical community. Neurological approaches to autism are popular, and many autistic adults embrace the neuroscripts that accompany these explanations, referring to people without autism as 'neurotypical'. By using the language of neuroscience, some autistic adults can anchor their experiences within a wider bioscience narrative. But the neurological challenges still fall within the biomedical model; the real difficulty is challenging this narrative while also transcending it, becoming captured by the paradigm with little room for manoeuvre. If Runswick-Cole were to avoid the language of autism—as many of us who might want to move away from labelling and categorising human beings might prefer—what kinds of reference points would need to be developed? This is the ongoing challenge for anyone committed to making sense of power relations, patriarchy and systems of violence that privilege perspectives of the dominant group over those in different kinds of marginal spaces.

1.2 Absence of the Social

It is not be surprising that those within a biomedical model of autism see the connections between autism and robotic machines more directly. In trying to model the absence of the social, psychologists and psychiatrists have looked to autism as a way to understand attachment and the enduring difficulties that come from failing to attach to persons. By contrast, robotics and AI scientists have looked to non-social forms of otherness—*the robotic machine*—as a way to gain insight into the experience of life on the autism spectrum. Mechanical metaphors of autism abound in the psychiatric and robot therapy literature: it is repeatedly asserted that there is a special affinity between children with autism and robots. A backlash has formed against these mechanical descriptions of what it means to have autism, however, involving parents challenging experts. For instance, parent Louise Milligan publicly challenged the view that children with autism have less empathy (The Guardian, Milligan July 2016).

One could say that (some) social scientists are abandoning the 'social' (Latour 2002) while technologists are absorbing it and extending it into the machines they create (Richardson 2015; Breazeal 2004). Entities created as an outcome of human action—art, computers, robots—were once seen as the repositories of their creator's agency. Now, however, these entities are viewed as agents in their own right; according to these approaches, the divide between humans and machines is artificial and is a categorical distinction worthy of dissolution.

This book brings together these two distinct fields of research that propose unique models of the 'social' and explores what happens when these fields develop a partnership. For the psychiatry and psychological science of autism, autism is a biomedical condition, shaped by heredity and genes (although no 'autism genes' have been found, strong evidence that autism runs in families is given in the literature) (Frith 1989). Current scholarship proposes a mixture of biomedical causes: excess of testosterone (Baron-Cohen 2005), heredity and genes, and a neurodevelopmental disorder of the brain (Baron-Cohen 1997, 2005). While established biomedical professionals may want to keep control over the narratives concerning the causes and treatments of autism, a range of alternatives (often dismissed as quackery) circulate as to the causes of autism. These include: attachment difficulties with primary caregivers; 'refrigerator parenting' (Kanner 1943; Bettelheim 1967); modern vaccines such as the rubella, mumps and measles (MMR) vaccine; diet; and pesticides (Grinker 2008; Godlee et al. 2011).

It was Leo Kanner who first popularised the idea that mothers (and fathers) of children with autism were emotionally distant. In the 1960s he went so far as to say that mothers of autistic children 'just happen to defrost enough to produce a child' (Time Magazine 1960). In Kanner's original paper he wrote about the parents of the children he saw as detached, highly educated men and women writing 'In the whole group, there are very few really warmhearted fathers and mothers' (Kanner 1943, p. 250). Attachment theories have been intimately tied to gender, particularly to 'mothering' and the impact of the mother on the child. Needless to say this perspective has caused considerable controversy over time and much distress among mothers of children with autism. Thus it is important to understand the way in which mothering has been culturally constructed to support gender roles between men and women in Euro-American societies.

In this book I explore a gendered theory of attachment, and the role of each parent in shaping a child. The mother role is traditionally performed by women (and still is to a large extent) and is expected to take primary responsibility for the child's day to day care. Fathers have been largely ignored because they were not primarily responsible for the care and development of 'affective' relations with their children. The way that children are raised remains a controversial field, with adult carers taking on responsibility for how their children develop in different ways.

Autism chiefly describes an altered experience of attachment, but children with autism may also display difficulties in cognition, memory, understanding humour, and imaginative play. Autism is a serious condition where children are unable to engage in reciprocal communication and interactions with even their closest and intimate caregivers, as well as in other types of relationships. It is important to underscore attachment as the primary difficulty that characterizes autism. This is not to say that children with autism cannot find ways to cope, live at ease with the attachment experience, or gain valuable support from caregivers, family and specialists. Some adults with autism have even suggested that attempts to make them become 'normal' significantly contributed to the difficulties they experienced as children, firstly in coming to terms with their own experiences, and then by having familial, educational and medical services attempt to shape their way of understanding, and being in, the world (Grinker 2008). The Autism Advocacy movement is an important voice in this narrative of autism. Adults with autism will also confirm that social interaction is difficult and that reading the social cues from others is frustrating and difficult (Grandin and Panek 2014; Baggs 2007 film). When it comes to difficulties in social interaction, both experts and those with autism are in some kind of agreement.

1.3 Pronouns and Relationality

Strathern's *The Gender of the Gift* (1988) proposed that Melanesians could not be gendered in the Euro-American sense as they had no sense of what it means to be an individual. Instead Melanesians were a composite of their relations with others—living and dead and person and thing. In other words, Melanesians had no sense of 'I' in the classic description of the individual. In Cartesian terms, the pronoun 'I' is tied directly with the consciousness of the 'I'; 'I think therefore I am'. This has led some to propose that the pronoun 'I' is tied to a lone subjectivity, an Island separated off from the mainland of relationship. I is a curious word as in English speaking languages it looks very similar to the figure for one, 1, and in Roman numerals, it is the same shape: I. There's also an archaic form of speech that is still used by the British monarchy to refer to the self as 'one'.

Stawarska (2009) in her book *Between I and You*, has shown how a separated subjectivity given by a lone 'I' is impossible: there can be no I without You; they are inseparable. In linguistic terms, the you precedes the I as the caregiver can utter the sounds you, before the child can express the words I. Stawarska continues: 'reduction of "the I" does not simply represent but rather produces a novel conception of the person as a repository of inner private events accessible by means of first-person insight…a conception of personhood modeled on the ideal of apodictic knowledge gained via intuitive insight leads to an exclusion of second person relatedness, and a forgetting of the inseparability of I and you' (p. 18).

The lone I is impossible: no human can survive without another human, there is no existence as a separated and self-contained individual. The tales of wolves, dogs and apes raising children have long suggested the possibility that human life could exist in some otherworldly realm, and that other species could take on the caregiver role for humans without jeopardizing their survival (Newton 2002). Accounts of animals fulfilling this caregiving role to humans in real life are rare. When such children are 'found' and reintegrated into the human world, those who discovered them are keen to note how the mannerisms or behaviours of these children differ from those raised by humans. Autism expert Uta Frith, a proponent of the genetic and neuroscientific model of autism, argues that examples of autism pre-exist Kanner's observational paper in the 1940s, referring to the Wild Boy of Aveyron, the case of Victor, found in a forest of central France (Frith 1989). The Wild Boy, she writes,

> …who appeared to be 12 years old, did not speak, did not respond to questions, did not even respond to noises made next to him. He had no clothes and his body was covered in scars. His whole appearance and range of behaviour seemed totally *asocial*. (p. 17 my emphasis)

In keeping with modern practice, Frith applies a retrospective clinical diagnosis to describe and make sense of the behaviours of a boy that was schooled in the wild. Applying retrospective diagnoses to people is a fashion in autism circles—everyone from Albert Einstein to Wolfgang Amadeus Mozart.

1.4 Machine Sociality

The development of robots for helping children with autism develop socially marks a turning point in autism studies. How did a condition like autism get picked up by a community of researchers to help them to develop robots? Why are robots able to help children with autism develop socially? The answer lies in the influence of the theories of autism by Simon Baron-Cohen and the appropriation of these ideas by the robotics community, which was already shaped by mechanistic theories of the social. I look at the main themes in turn.

Theory of Mind and Mindblindness

The Theory of Mind (ToM) is a theory to describe how humans can coexist without direct access to the precise content of other agents' minds. It is a predictive theory: we imagine the thoughts and feelings of other persons (and animals) by making guesses and assumptions about other agents. Baron-Cohen used the term 'mindblind' to describe children with autism, and argued they lack a ToM (Baron-Cohen 1997). Moreover, as non-social beings, robots also lack a ToM, but Scassellati (Scassellati 2002) proposed that robots must implement some social awareness and develop a ToM.

Extreme Male Brain and Systemising

The EMB is a theory of sex differences between males and females developed by Baron-Cohen to explain preferences of males and females (Baron-Cohen 2004). According to Baron-Cohen, males naturally 'systemise'—developing systems and finding out what mechanisms underlie systems. Females, by contrast, naturally 'empathise'—this involves taking care of others and putting others' needs before their own. These natural differences, according to Baron-Cohen, account for the genius of men in science, technology, engineering and mathematics (Baron-Cohen et al. 1997, 2009b; Baron-Cohen and Wheelwright 2004; Baron-Cohen 2004, 2011). The observation that children with autism initiate fewer social interactions than children with typical development (or neurotypical children) along with the higher proportion of males diagnosed with autism are all used as evidence to support EMB theory.

The claim that children with autism prefer robots (as they are mechanical objects) over other kinds of entities, especially over other persons, drives the philosophy of robot therapy. By ascribing social qualities to machines, the roboticists speculate that robot therapy can provide an intermediary realm, somewhere between the typical social world and the non-social machine world. In anthropomorphising robots, these roboticists change the ontological status of robots so that the latter are presented as not merely inanimate artefacts but 'something in between'. Undergirding this notion about the suitability of robots for children with autism is an understanding of sociality as mechanical, inspired by the machine. A new definition of the machine to illustrate this worldview is given below:

> *The machine is the parts: no matter what is added to or taken from it, the machine can never be assembled to make up the interpersonal whole.*

In the above definition, the machine is a set of ideas and practices that carve up the human person into parts. But the machine is more than that. The machine is a practice, a theorising process and entity that is in parts, and can never be assembled to make a whole that is plural. This plurality is an outcome of species-specific sociality and attachment. In mechanical sociality, humans have no essential 'social relationships' that make them different from artefacts that also can have 'social relationships'.

Machines, if given anthropomorphic or social qualities, are social. In contemporary theorising about humanoid robotics and autism, physical machines as well as people on the autism spectrum are imagined as having too few parts: a deficit of parts (empathy); or having an *excess* of parts (systemising). The robot and child with autism then coexist in an analogical universe. A *Principle of Equivalence* operates where things of different values can become commensurate. Equivalence works best when thinking about commodities. For instance, Marx's theory of value uses an algebraic formulation to describe how different types of commodities can be made equivalent:

> X commodity a = y commodity b, x commodity a is worth y commodity b. (1974, p. 55)

The principle of equivalence is not only about bringing different things into relation with each other as commensurate values but also about ignoring their intrinsic aspects. A coat is different from a pair of shoes, but as values can become commensurate, one can be traded against the other. Analogies and metaphors, unlike commodities, have different properties; they allow things to be read, as something other, without becoming the other. In the field of robots for autism, autistic meanings flow into 'social' robots and back again, but is the analogy between 'social' robots and autism something we should be concerned about? In some psychological theories of autism, autistic persons are non-social, preferring objects and systems to other people. According to these widespread ideas, autistic persons are said to lack empathy, a crucial social experience that is fundamental to human co-experience. Autism expert Simon Baron-Cohen had this to say about empathy: 'empathy is without question an important ability. It allows us to tune into how somene else is feeling, or what they might be thinking… In short, empathy allows us to interact effectively in the social world' (2011, p. 163). In the reverse direction, theories of autism flow from the psychiatrists' models into clinical engagements with real people, texts, seminars, research, diagnoses, conversations and labels. The biomedically informed psychological sciences and robotics have a lot in common, as shown by their increasing collaboration around robots and autism.

As anthropologists, we reflect on the methodological practice of the discipline, as well as attempting to write about the lifeworlds of those with whom we engage. As Strathern (1988) explains, a book is an artefact—a construction that is tied by its own workings and logic. In this text I write about living persons, real children and adults that I met in person, with whom I interacted, and I am responsible for translating and making sense of these encounters. It is a huge responsibility that requires a deep respect for those who share their lifeworlds with us. As Goodley so accurately wrote: 'Any debate, analytical intervention or rhetorical interrogation of the many faces of autism has the potential to upset somebody' (2016). Therefore, more than an ethnographic account of the use of robots for children with autism, my book is a theoretical contribution to anthropological theorising about humans and nonhumans and the relationships between them, or non-relationships between them. If human relationships are marked by co-experience and mutuality, then machines do not experience nor encounter humans in these ways—even though social roboticists claim they do.

Chapter 2 explores how the robots are imagined and put to use to help children with autism develop social behaviours and skills. I compare those models of autism that emphasize the autistic person as preferring things and machines over other people and which consider autism as part of the empathy-deficit model. I introduce alternative ways of thinking about autism that attempt to restore the importance of sociality and social relationships for autistic persons.

Chapter 3 is an ethnographic account of an experiment conducted at a specialist school for children with autism. The students of the school were participants in a small experiment to study the effectiveness of a humanoid robot for developing social skills for children with autism. The experiment failed to show any difference between children who interacted with a robot or a toy crane for the duration of the experiment. The results were shelved and never published as are many experiments that do not show positive results. This chapter then explores the politics of the 'experiment' as a way of knowing about the world, or keeping knowledge hidden.

Chapter 4 explores the mythology of Echo and Narcissus and how this mythology informs studies of autism. This chapter explores how childhood studies are situated within an egocentric individualistic paradigm. The orthodoxy of this paradigm has created myths about childhood as a non-relational, selfish and non-empathetic state. Again myths that connect autism with detached or 'narcissistic' states are explored and challenged. The use of the narcissistic myth is disconnected from the story of Echo, which I reunite and show they are importantly connected to each other. Social communication takes place through a series of social behaviours that include turn-taking, reciprocal interactions, pronoun use, joint attention and imitation. In light of the way in which echolalia is a feature of autism, I look at how pronoun-reversal (the taking of opposite speech roles: You in place of I) and how these ways of understanding interpersonal experience is inflected through autism and wider notions of the individual in capitalism.

In Chap. 5 I look at attachment theory and the social history of parenting and motherhood. I explore the role of parenting and attachment as important to unfolding this narrative. Attachment theory is the study of relationship developed as a research area in the 1940s. Developed by John Bowlby, a psychoanalyst, attachment theory has been developed in a multitude of ways. I use attachment theory to discuss interpersonal relationships. There are many critics of Bowlby's work, particularly from feminists who see his emphasis on female caregiving as problematic, ignoring the role of fathers or other caregivers in a child's life. This chapter will explore these themes.

In Chap. 6 I explore the etymology of the term 'machine'—from the ancient Greek mēkhos, meaning 'contrivance', which pertains to the contriving of something with skill, inventiveness or deceit. Machine as contrivance is a device or a map and not the territory. Central to imagination is analogical thinking, which allows one thing to stand in place for something else without becoming the thing it stands in for. This type of imaginative thinking is said to be deficient in autistic persons, much like an understanding of metaphor, empathy or humour. But could this type of imaginative thinking be deficient in wider society where analogies between humans and machines are dissolving in place of identification between human and machine? Drawing on the way in which autism is understood through the 'deficit' or 'excess' model, I explore how the robot's interventions in the child's life are imagined to bring about stability through equivalence. This principle of equivalence between children with autism and social robots is underscored by mathematical models built from commercial models of the human.

Chapter 7 explores an exciting new field of Critical Autism Studies (CAS), a movement of adults with autism, parents, psychotherapists, psychologists, psychiatrists, disabled activists and social scientists creating a new narrative around autism and challenging existing orthodoxy in autism studies that turns autistic persons into 'brains' (neuroscience) or 'things' (a set of diagnostic categories of mental disorders). This movement refocuses our attention on the humanity and interpersonal relationships of people with autism and the multitude of relationships they encounter through various health services. Robot therapy is built on problematic foundations that view people with autism as lacking in empathy or meaningful sociality. CAS could provide the template for breaking away from the dehumanising descriptions of autistic persons towards an integrative humanistic paradigm.

Chapter 8, the conclusion, proposes an alternative model to the egocentric individualistic model of autism which has dissolved the boundaries between humans and machines. Egocentric individualists are challenged by anti-humanists who want to reject human essentialism in favour of poly-perverse monsters, cyborgs and ontologies that equate persons and things. These two modes of thought arise out of each other, for cyborgs would have nothing to dissolve if there was not a radical ontological break driven by egocentric individualism. Instead I propose a humanistic I-you interrelatedness to make sense of human beings, and the set of those human characteristics that psychologists frame and label to produce 'autism'. As I-you interrelatedness approaches privilege the centrality of interpersonal experience as a dual back and forth, the humanistic approach does not foreclose a you in favour of an I or an I in favour of a you. Humanistic I-you approaches allow us to explore what is possible with robots as tools to help humanity without following the course I describe in this book of becoming analogous with them and merging into machines.

Bibliography

Ainsworth, Mary D. Salter. 1978. The Bowlby-Ainsworth Attachment Theory. *Behavioral and Brain Sciences* 1 (03): 436–438.

American Psychiatric Association. 2013a. *Diagnostic and Statistical Manual of Mental Disorders (DSM-5®)*. American Psychiatric Pub.

American Psychiatric Association. 1994. DSM-IV: Diagnostic and Statisticalmanual. Washington, DC: American Psychiatric Association.

American Psychiatric Association, 2013. Diagnostic and statistical manual of mental disorders (DSM-5®). American Psychiatric Pub.

Bagatell, Nancy. 2010. From Cure to Community: Transforming Notions of Autism. *Ethos* 38 (1): 33–55.

Baron-Cohen, S. 1997. *Mindblindness: An Essay on Autism and Theory of Mind*. Cambridge: MIT Press.

———. 2004. *The Essential Difference*. London: Penguin.

———. 2011. *Zero Degrees of Empathy: A New Theory of Human Cruelty*. Vol. 30. London: Penguin.

Baron-Cohen, Simon, and Jessica Hammer. 1997. Is Autism an Extreme Form of the "Male Brain"? *Advances in Infancy Research* 11: 193–218.

Baron-Cohen, Simon, and Sally Wheelwright. 2004. The Empathy Quotient: An Investigation of Adults with Asperger Syndrome or High Functioning Autism, and Normal Sex Differences. *Journal of Autism and Developmental Disorders* 34 (2): 163–175.

Baron-Cohen, Simon, Sally Wheelwright, Carol Stott, Patrick Bolton, and Ian Goodyer. 1997. Is There a Link Between Engineering and Autism? *Autism-London* 1: 101–109.

Baron-Cohen, Simon, Patrick Bolton, Sally Wheelwright, Victoria Scahill, Liz Short, Genevieve Mead, and Alex Smith. 1998. Autism Occurs More Often in Families of Physicists, Engineers, and Mathematicians. *Autism* 2 (3): 296–301.

Baron-Cohen, Simon, Ofer Golan, and Emma Ashwin. 2009a. Can Emotion Recognition Be Taught to Children with Autism Spectrum Conditions? *Philosophical Transactions: Biological Sciences* 364 (1535): 3567–3574.

Baron-Cohen, Simon, Ofer Golan, Emma Chapman, and Yael Granader. 2009b. Transported to a World of Emotion. *McGill Journal of Medicine* 12 (2): 78–81.

Baron-Cohen, S., E. Ashwin, C. Ashwin, T. Tavassoli, and B. Chakrabarti. 2009c. Talent in Autism: Hyper-Systemizing, Hyper-Attention to Detail and Sensory Hypersensitivity. *Philosophical Transactions of the Royal Society of London: Series B, Biological Sciences* 364 (1522): 1377–1383.

Baudrillard, Jean. 1994. *Simulacra and Simulation*. Ann Arbor: University of Michigan Press.

Bernays, Edward L. 2005 [1928]. *Propaganda. With an Introduction by Mark Crispin Miller*. Brooklyn: Ig Publishing.

Bettelheim, Bruno. 1959a. Feral Children and Autistic Children. *American Journal of Sociology* 64: 455–467.

———. 1959b. Joey: A Mechanical Boy. *Scientific American* 200 (3): 116–120.

———. 1967. *Empty Fortress*. New York/London: Free Press/Collier-Macmillan.

Bibles, Collins Anglicised ESV. 2008. *The Holy Bible: English Standard Version*. London: HarperCollins.

Billington, T. 2016. Critical Autism and Critical Neuroscience: Towards a Science of Research and Practice. In *Re-thinking Autism: Diagnosis, Identity and Equality*, ed. K. Runswick-Cole, R. Mallett, and S. Timimi. London/Philadelphia: Jessica Kingsley Publishers.

Bird, Geoffrey, Jane Leighton, Clare Press, and Cecilia Heyes. 2007. Intact Automatic Imitation of Human and Robot Actions in Autism Spectrum Disorders. *Proceedings: Biological Sciences* 274 (1628): 3027–3031.

Blacher, Jan, and Lisa Christensen. 2011. Sowing the Seeds of the Autism Field: Leo Kanner (1943). *Intellectual and Developmental Disabilities* 49 (3): 172–191.

Boellstorff, Tom. 2015. *Coming of Age in Second Life: An Anthropologist Explores the Virtually Human*. Princeton University Press.

Bourdieu, P. 1992. The Logic of Practice. Cambridge: Polity Press. Paperback edition.

———. 2011. The Forms of Capital (1986). *Cultural Theory: An Anthology*: 81–93.

Bowlby, John. 1973. *Attachment and Loss: Anxiety and Anger*. Vol. 95. London: Hogarth Press.

———. 1980. *Attachment and Loss: Sadness and Depression*. Vol. 109. London: Penguin.

———. 1982. *Attachment and Loss*. 2nd ed. Vol. 79. London: Hogarth Press.

———. 2008, New York. *Attachment*. Basic Books.

Breazeal, Cynthia L. 2004. *Designing Sociable Robots*. Cambridge, MA/London: MIT Press.

Breazeal, Cynthia. 2009. Role of Expressive Behaviour for Robots That Learn from People. *Philosophical Transactions: Biological Sciences* 364 (1535): 3527–3538.

Broadbent, Elizabeth, Rebecca Stafford, and Bruce MacDonald. 2009. Acceptance of Healthcare Robots for the Older Population: Review and Future Directions. *International Journal of Social Robotics* 1 (4): 319.

Brooks, R. 2002. *Flesh and Machines: How Robots Will Change Us*. New York: Patheon.

Brownlow, Charlotte, and Lindsay O'Dell. 2009. Challenging Understandings of "Theory of Mind": A Brief Report. *Intellectual and Developmental Disabilities* 47 (6): 473–478.

Buber, M. 1937. *I and Thou*. Trans: R.G. Smith. Edinburgh, Clark.

Candland, Douglas K. 1995. *Feral Children and Clever Animals: Reflections on Human Nature*. New York: Oxford University Press on Demand.

Casson, John. 2004. *Drama, Psychotherapy and Psychosis: Dramatherapy and Psychodrama with People Who Hear Voices*. New York: Routledge.

Castelli, F. 2005. Understanding Emotions from Standardized Facial Expressions in Autism and Normal Development. *Autism: The International Journal of Research and Practice* 9 (4): 428–449.

Castells, M. 2000. *The Rise of the Network Society: Economy, Society and Culture v.1: The Information Age: Economy, Society and Culture*. Vol. 1. Oxford, UK/ Malden, MA: Wiley-Blackwell.

Coeckelbergh, Mark, Cristina Pop, Ramona Simut, Andreea Peca, Sebastian Pintea, Daniel David, and Bram Vanderborght. 2016. A Survey of Expectations About the Role of Robots in Robot-Assisted Therapy for Children with ASD: Ethical Acceptability, Trust, Sociability, Appearance, and Attachment. *Science and Engineering Ethics* 22 (1): 47–65.

Cohen-Rottenberg, Rachel. 2012. Deconstructing Autism as an Empathy Disorder: A Literature Review. Autism and Empathy. Online http://www.disabilityandrepresentation.com/deconstructing-autism-as-an-empathy-disordera-literature-review/

Collins, H.M. 1998. Socialness and the Undersocialized Conception of Society. *Science, Technology, & Human Values* 23 (4): 494–516.

Collins, G. 2016. Does a Diagnosis of ASD Help Us to Help a Person with Intellectual Disabilities? In *Re-thinking Autism: Diagnosis, Identity and Equality*, ed. K. Runswick-Cole, R. Mallett, and S. Timimi. London/ Philadelphia: Jessica Kingsley Publishers.

Costescu, Cristina A., and Daniel O. David. 2014. Attitudes Toward Using Social Robots in Psychotherapy. *Erdelyi Pszichologiai Szemle= Transylvanian Journal of Psychology* 15 (1): 3.

Crittenden, P.M. and Ainsworth, M.D., 1989. Chapter 14 Child maltreatment and attachment theory (pp. 432–463). In ed. Dante Cicchetti and Vicki Carlson. Child Maltreatment: Theory and Research on the Causes and Consequences of child abuse and neglect. London, England: Cambridge University Press.

Cromer, Richard. 1974. The Development of Language and Cognition: The Cognition Hypothesis. In *New Perspectives in Child Development*, ed. B. Foss, 184–252. Harmondsworth: Penguin.

Danforth, Scot, and Srikala Naraian. 2007. Use of the Machine Metaphor Within Autism Research. *Journal of Developmental and Physical Disabilities* 19 (3): 273–290.

Dautenhahn, Kerstin. 2007. Socially Intelligent Robots: Dimensions of Human-Robot Interaction. *Philosophical Transactions: Biological Sciences* 362 (1480): 679–704.

Dautenhahn, Kerstin, and Iain Werry. 2004. Towards Interactive Robots in Autism Therapy: Background, Motivation and Challenges. *Pragmatics & Cognition* 12 (1): 1–35.

Davis, K. 2016. How Rude? Autism as a Study in Ability. In *Re-thinking Autism: Diagnosis, Identity and Equality*, ed. K. Runswick-Cole, R. Mallett, and S. Timimi. London/Philadelphia: Jessica Kingsley Publishers.

de Graaf, M.M.A. 2016. An Ethical Evaluation of Human–Robot Relationships. *International Journal of Social Robotics* 8 (4): 589–598.

de La Mettrie, Julien Offray, and I.I. Frederick. 1912. *Man a Machine*. Chicago: Open Court Publishing Company.

de Solla Price, Derek J. 1964. Automata and the Origins of Mechanism and Mechanistic Philosophy. *Technology and Culture*: 9–23.

Delvaux, M., 2016. DRAFT REPORT with recommendations to the Commission on Civil Law Rules on Robotics (2015/2103 (INL)). Committee on Legal Affairs. European Parliament, PR\1095387EN. doc, PE582. 443v01–00.

Dennis, Wayne. 1941. The Significance of Feral Man. *The American Journal of Psychology* 54 (3): 425–432.

Diehl, Joshua J., Lauren M. Schmitt, Michael Villano, and Charles R. Crowell. 2012. The Clinical Use of Robots for Individuals with Autism Spectrum Disorders: A Critical Review. *Research in Autism Spectrum Disorders* 6 (1): 249–262.

Dourish, Paul, and Genevieve Bell. 2011. *Divining a Digital Future: Mess and Mythology in Ubiquitous Computing*. Cambridge, MA: MIT Press.

Dworkin, Andrea, and Marvin Heiferman. 1981. *Pornography: Men Possessing Women*. Vol. 1. London: Women's Press.

Dworkin, Andrea, and Catharine A. MacKinnon. 1988. *Pornography and Civil Rights: A New Day for Women's Equality*. Minneapolis: Organizing Against Pornography.

Engels, Friedrich. 2010. *The Origin of the Family, Private Property and the State. With an Introduction by Tristram Hunt*. London: Penguin Classics.

Esteban, Pablo Gómez, Hoang-Long Cao, Albert De Beir, Greet Van de Perre, Dirk Lefeber, and Bram Vanderborght. 2016. A Multilayer Reactive System for Robots Interacting with Children with Autism. arXiv Preprint arXiv:1606.03875.

Evans, Bonnie. 2013. How Autism Became Autism: The Radical Transformation of a Central Concept of Child Development in Britain. *History of the Human Sciences* 26 (3): 3–31.

Eyal, Gil. 2010. *The Autism Matrix*. Cambridge: Polity.

Feil-Seifer, David and Maja J. Mataric. 2005. Defining Socially Assistive Robotics. *IEEE*. Chicago, Il, USA.

Fine, Cordelia. 2005. *Delusions of Gender: The Real Science Behind Sex Differences*. New York: Icon Books Ltd.

Foucault, Michel, and Jay Miskowiec. 1986. Of Other Spaces. *Diacritics* 16 (1): 22–27.

Freud, Sigmund, and James Strachey. 1950. *Totem and Taboo: Some Points of Agreement Between the Mental Lives of Savages and Neurotics*. London: Routledge and Kegan Paul.

Frith, U. 1989. Autism: explaining the enigma. Oxford: Basil Blackwell.

———. 1991. *Autism and Asperger Syndrome*. Cambridge/New York: Cambridge University Press.

Gell, Alfred. 1998. *Art and Agency: An Anthropological Theory*. Oxford: Clarendon Press.

Gershon, Ilana. 2010. *The Breakup 2.0: Disconnecting Over New Media*. Ithaca: Cornell University.

Gilligan, Carol. 1982. *In a Different Voice*. Cambridge, MA/London: Harvard University Press.

Godlee, F., J. Smith, and H. Marcovitch. 2011. Wakefield's Article Linking MMR Vaccine and Autism Was Fraudulent. *BMJ: British Medical Journal* 342 (7788): 64–66.

Gopnik, Alison. 1993. Theories and Illusions. *Behavioral and Brain Sciences* 16 (01): 90–100.

Gould, Judith, and Jacqui Ashton-Smith. 2011. Missed Diagnosis or Misdiagnosis? Girls and Women on the Autism Spectrum. *Good Autism Practice (GAP)* 12 (1): 34–41.

Goodley, D. 2016. Autism and the Human. In *Re-thinking Autism: Diagnosis, Identity and Equality*, ed. K. Runswick-Cole, R. Mallett, and S. Timimi, 146–158. London/Philadelphia: Jessica Kingsley Publishers.

Grandin, T., and R. Panek. 2014. *The Autistic Brain: Exploring the Strength of a Different Kind of Mind*. London/Sydney: Rider Books.

Grinker, Roy Richard. 2008. *Unstrange Minds: Remapping the World of Autism*. New York: Basic Books.

———. 2010. Commentary: On Being Autistic, and Social. *Ethos* 38 (1): 172–178.

Grisaffe, Douglas B., and Hieu P. Nguyen. 2011. Antecedents of Emotional Attachment to Brands. *Journal of Business Research* 64 (10): 1052–1059.

Hacking, I. 2009. Autistic Autobiography. *Philosophical Transactions of the Royal Society of London: Series B, Biological Sciences* 364 (1522): 1467–1473.

Hampson, Norman. 1990. *The Enlightenment*. Vol. 4. Harmondsworth: Penguin.

Happé, Francesca, and Uta Frith. 2010. *Autism and Talent*. Vol. 364. Oxford: Oxford University Press.

Haraway, Donna Jeanne. 1991. *Simians, Cyborgs and Women: The Reinvention of Nature*. London: Free Association Books.

———. 2003. *The Companion Species Manifesto: Dogs, People, and Significant Otherness*. Vol. 1. Chicago: Prickly Paradigm Press.

Harlow, Harry F. 1958. The Nature of Love. *American Psychologist* 13 (12): 673.

Harlow, Harry F., and Robert R. Zimmermann. 1958. The Development of Affectional Responses in Infant Monkeys. *Proceedings of the American Philosophical Society* 102 (5): 501–509.

Hashim, R., and H. Yussof. 2017. Humanizing Humanoids Towards Social Inclusiveness for Children with Autism. *Procedia Computer Science* 105 (4): 359–364.

Hassall, R. 2016. Does Everyone with Autism Have the Same Underlying Condition? In *Re-thinking Autism: Diagnosis, Identity and Equality*, ed. K. Runswick-Cole, R. Mallett, and S. Timimi. London/Philadelphia: Jessica Kingsley Publishers.

Hasse, Cathrine. 2002. Gender Diversity in Play with Physics: The Problem of Premises for Participation in Activities. *Mind, Culture, and Activity* 9 (4): 250–269.

Haydon-Laurelut, M. 2016. Critical Systemic Therapy: Autism Stories and Disabled People with Learning Difficulties. In *Re-thinking Autism: Diagnosis, Identity and Equality*, ed. K. Runswick-Cole, R. Mallett, and S. Timimi, 221–238. London/Philadelphia: Jessica Kingsley Publishers.

Hazan, Cindy, and Phillip Shaver. 1987. Romantic Love Conceptualized as an Attachment Process. *Journal of Personality and Social Psychology* 52 (3): 511.

Helmreich, Stefan. 1998. *Silicon Second Nature: Culturing Artificial Life in a Digital World, Updated with a New Preface*. Berkeley: University of California Press.

Hobson, R. Peter. 1995. *Autism and the Development of Mind*. Hove/Hillsdale: Laurence Erlbaum Associates, Publishers.

Hodge, N. 2016. Schools Without Labels. In *Re-thinking Autism: Diagnosis, Identity and Equality*, ed. K. Runswick-Cole, R. Mallett, and S. Timimi. London/Philadelphia: Jessica Kingsley Publishers.

Hollin, Gregory. 2014. Autism, Sociality, and Human Nature. *Somatosphere.net*. Online: http://somatosphere.net/2014/06/autism-sociality-and-human-nature.html

Holmes, Jeremy. 1993. *John Bowlby and Attachment Theory*. London: Routledge.

Horlin, Chiara, Marita Falkmer, Richard Parsons, Matthew A. Albrecht, and Torbjorn Falkmer. 2014. The Cost of Autism Spectrum Disorders. *PLoS One* 9 (9): e106552.

Huber, A., A Weiss, and M Rauhala. 2016. The Ethical Risk of Attachment: How to Identify, Investigate and Predict Potential Ethical Risks in the Development of Social Companion Robots. Conference paper. HRI '16 The Eleventh ACM/IEEE International Conference on Human Robot Interaction. Pp. 367–374.

Huijnen, Claire A.G.J., Monique A.S. Lexis, Rianne Jansens, and Luc P. Witte. 2016. Mapping Robots to Therapy and Educational Objectives for Children with Autism Spectrum Disorder. *Journal of Autism and Developmental Disorders* 46 (6): 2100–2114.

Kanner, L., 1943. Autistic disturbances of affective contact. Nervous child, 2(3), pp.217–250.

Kanner, L. 1960. Do Behavioural Symptoms Always Indicate Psychopathology? *Journal of Child Psychology and Psychiatry* 1 (1): 17–25.

Kaufman, Sharon R. 2010. Regarding the Rise in Autism: Vaccine Safety Doubt, Conditions of Inquiry, and the Shape of Freedom. *Ethos* 38 (1): 8–32.

Keller, Evelyn Fox. 2009. *The Century of the Gene*. Cambridge, MA: Harvard University Press.

Kelty, Christopher. 2005. Geeks, Social Imaginaries, and Recursive Publics. *Cultural Anthropology* 20 (2): 185–214.

Kim, E.S.W. 2013. Robots for Social Skills Therapy in Autism: Evidence and Designs Toward Clinical Utility. Doctoral Dissertation, Yale University, New Haven. Online at ACM Digital Library: https://dl.acm.org/citation.cfm?id=2604248

Kim, Joseph A., Peter Szatmari, Susan E. Bryson, David L. Streiner, and Freda J. Wilson. 2000. The Prevalence of Anxiety and Mood Problems Among Children with Autism and Asperger Syndrome. *Autism* 4 (2): 117–132.

Koehn, Daryl. 1998. *Rethinking Feminist Ethics: Care, Trust and Empathy*. London/New York: Routledge.

Kozima, H., C. Nakagawa, and Y. Yasuda. 2005, August. Interactive Robots for Communication-Care: A Case-Study in Autism Therapy. In *Robot and Human Interactive Communication*. ROMAN 2005. IEEE International Workshop on (pp. 341–346). IEEE.

Kurzweil, Ray. 2000. *The Age of Spiritual Machines: When Computers Exceed Human Intelligence*. New York: Penguin.

———. 2005. *The Singularity Is Near: When Humans Transcend Biology*. New York: Penguin.

Lakoff, George. 1987. *Women, Fire, and Dangerous Things: What Categories Reveal About the Mind*. Chicago: The University of Chicago Press.

Lakoff, George, and Mark Johnson. 2008. *Metaphors We Live By*. Chicago: University of Chicago Press.

Lasch, Christopher. 1991. *The Culture of Narcissism: American Life in an Age of Diminishing Expectations*. New York/London: W.W. Norton & Company.

Latif, S. 2016. The Ethics and Consequences of Making Autism Spectrum Disorder. In *Re-thinking Autism: Diagnosis, Identity and Equality*, ed. K. Runswick-Cole, R. Mallett, and S. Timimi. London/Philadelphia: Jessica Kingsley Publishers.

Latour, B. 2012. Gabriel Tarde and the End of the Social. In *The Social in Question. New Bearings in History and the Social Sciences*, ed. Patrick Joyce, 117–132. London: Routledge.

———. 2005. *Reassembling the Social: An Introduction to Actor-Network-Theory*. Oxford/New York: Oxford University Press.

———. 2012. *We Have Never Been Modern*. Harvard University Press.

Lee, Anthony, R. Peter Hobson, and Shulamuth Chiat. 1994. I, You, Me, and Autism: An Experimental Study. *Journal of Autism and Developmental Disorders* 24 (2): 155–176.

Li, C., Q. Jia, and Y. Feng. 2016, August. Human-Robot Interactoin Design for Robot-Assisted Intervention for Children with Autism Based on ES Theory. In *Intelligent Human-Machine Systems and Cybernetics (IHMSC)*, 8th International Conference on (Vol. 2, pp. 320–324). IEEE.

Liss, Miriam, and Mindy J. Erchull. 2012. Feminism and Attachment Parenting: Attitudes, Stereotypes, and Misperceptions. *Sex Roles* 67 (3–4): 131–142.

Loh, Alvin, Teesta Soman, Jessica Brian, Susan E. Bryson, Wendy Roberts, Peter Szatmari, Isabel M. Smith, and Lonnie Zwaigenbaum. 2007. Stereotyped Motor Behaviors Associated with Autism in High-Risk Infants: A Pilot Videotape Analysis of a Sibling Sample. *Journal of Autism and Developmental Disorders* 37 (1): 25–36.

Mallet, R., and K. Runswick-Cole. 2016. The Commodification of Autism: What's at Stake? In *Re-thinking Autism: Diagnosis, Identity and Equality*, ed. K. Runswick-Cole, R. Mallett, and S. Timimi. London/Philadelphia: Jessica Kingsley Publishers.

Margaret, Davies, and Ngaire Naffine. 2001. Are Persons Property. *Legal Debates About Property and Personality* 54.

Martin, Emily. 1991. The Egg and the Sperm: How Science Has Constructed a Romance Based on Stereotypical Male-Female Roles. *Signs: Journal of Women in Culture and Society* 16 (3): 485–501.

Marx, K., and F. Engels. 1974. *Capital: A Critique of Political Economy. Book 1, The Process of Production of Capital.* London: Lawrence & Wishart.

McCarthy, Anjanie, Kang Lee, Shoji Itakura, and Darwin W. Muir. 2006. Cultural Display Rules Drive Eye Gaze During Thinking. *Journal of Cross-Cultural Psychology* 37 (6): 717–722.

McGuire, A. 2016. Life Without Autism: A Cultural Logic of Violence. In *Re-thinking Autism: Diagnosis, Identity and Equality*, ed. K. Runswick-Cole, R. Mallett, and S. Timimi. London/Philadelphia: Jessica Kingsley Publishers.

Miller, Alice. 1987. *The Drama of Being a Child: And the Search for the True Self.* London: Virago.

Milton, D.E. 2014. Autistic Expertise: A Critical Reflection on the Production of Knowledge in Autism Studies. *Autism: The International Journal of Research and Practice* 18 (7): 794–802.

Muskett, T. 2016. Examining Language and Communication in Autism Spectrum Disorder – In Context. In *Re-thinking Autism: Diagnosis, Identity and Equality*, ed. K. Runswick-Cole, R. Mallett, and S. Timimi. London/Philadelphia: Jessica Kingsley Publishers.

Nadel, Jacqueline, Arnaud Revel, Pierre Andry, and Philippe Gaussier. 2004. Toward Communication: First Imitations in Infants, Low-Functioning Children with Autism and Robots. *Interaction Studies* 5 (1): 45–74.

Newton, Michael. 2002. *Savage Boys and Wild Girls: A History of Feral Children.* London: Faber and Faber.

O'Dell, Lindsay, Hanna Bertilsdotter Rosqvist, Francisco Ortega, Charlotte Brownlow, and Michael Orsini. 2016. Critical Autism Studies: Exploring Epistemic Dialogues and Intersections, Challenging Dominant Understandings of Autism. *Disability & Society* 31 (2): 166–179.

Ochs, Elinor, and Olga Solomon. 2010. Autistic Sociality. *Ethos* 38 (1): 69–92. https://doi.org/10.1111/j.1548-1352.2009.01082.x.

Ochs, Elinor, Tamar Kremer-Sadlik, Karen Gainer Sirota, and Olga Solomon. 2004. Autism and the Social World: An Anthropological Perspective. *Discourse Studies* 6 (2): 147–183.

Ovid and Mary M. Innes. 1955. *The Metamorphoses of Ovid*. Harmondsworth: Penguin.

Pearce, Celia, Tom Boellstorff, and Bonnie A. Nardi. 2011. *Communities of Play: Emergent Cultures in Multiplayer Games and Virtual Worlds*. Cambridge, MA: MIT Press.

Peca, Andreea. 2016. Robot Enhanced Therapy for Children with Autism Disorders: Measuring Ethical Acceptability. *IEEE Technology and Society Magazine* 35 (2): 54–66.

Peca, Andreea, Ramona Simut, Sebastian Pintea, Cristina Costescu, and Bram Vanderborght. 2014. How Do Typically Developing Children and Children with Autism Perceive Different Social Robots? *Computers in Human Behavior* 41: 268–277.

Piaget, Jean. 1964. Part I: Cognitive Development in Children: Piaget Development and Learning. *Journal of Research in Science Teaching* 2 (3): 176–186.

———. 2013. *The Construction of Reality in the Child*. Vol. 82. London/New York: Routledge.

Pinch, Trevor J., and Wiebe E. Bijker. 1984. The Social Construction of Facts and Artefacts: Or How the Sociology of Science and the Sociology of Technology Might Benefit Each Other. *Social Studies of Science* 14 (3): 399–441.

Pino, M., M. Boulay, F. Jouen, and A.-S. Rigaud. 2015. "Are We Ready for Robots That Care for Us?" Attitudes and Opinions of Older Adults Toward Socially Assistive Robots. *Frontiers in Aging Neuroscience* 7: 141.

Pop, Cristina Anamaria, Ramona Simut, Sebastian Pintea, Jelle Saldien, Alina Rusu, Daniel David, Johan Vanderfaeillie, Dirk Lefeber, and Bram Vanderborght. 2013. Can the Social Robot Probo Help Children with Autism to Identify Situation-Based Emotions? A Series of Single Case Experiments. *International Journal of Humanoid Robotics* 10 (03): 1350025.

Pop, Cristina A., Sebastian Pintea, Bram Vanderborght, and Daniel O. David. 2014. Enhancing Play Skills, Engagement and Social Skills in a Play Task in ASD Children By Using Robot-Based Interventions. A Pilot Study. *Interaction Studies* 15 (2): 292–320.

Richardson, Kathleen. 2010. Disabling as Mimesis and Alterity: Making Humanoid Robots at the Massachusetts Institute of Technology. *Etnofoor* 22 (1): 75–90.

———. 2015. *An Anthropology of Robots and AI: Annihilation Anxiety and Machines*. Vol. 20. New York: Routledge.

———. 2016. Are Sex Robots as Bad as Killing Robots? In *What Social Robots Can and Should Do: Proceedings of Robophilosophy 2016/TRANSOR*, ed. J. Seibt, M. Norskov, and S. Schack, 290, pp. 27–30.

———. 2016a. Sex Robot Matters: Slavery, the Prostituted, and the Rights of Machines. *IEEE Technology and Society Magazine* 35 (2): 46–53.

Robins, Ben, Kerstin Dautenhahn, R. Te Boekhorst, and Aude Billard. 2005. Robotic Assistants in Therapy and Education of Children with Autism: Can a Small Humanoid Robot Help Encourage Social Interaction Skills? *Universal Access in the Information Society* 4 (2): 105–120.

Robins, Ben, Kerstin Dautenhahn, and Janek Dubowski. 2006. Does Appearance Matter in the Interaction of Children with Autism with a Humanoid Robot? *Interaction Studies* 7 (3): 509–542.

Robinson, J. 2014. Participatory Research with Adults with Asperger's Syndrome: Using Spatial Analysis to Explore How They Make Sense of Their Experience. Doctoral Dissertation, De Montfort University, Leicester.

Rosqvist, Hanna Bertilsdotter, and Charlotte Brownlow. 2015. "What's the Point of having Friends?": Reformulating Notions of the Meaning of Friends and Friendship Among Autistic People. *Disability Studies Quarterly* 35 (4).

Rousseau, Jean-Jacques. 2000. *Discourse on the Origin of Inequality*. Oxford: Oxford University Press.

Roy, Souvik, Abhilash Ponnam, and Santanu Mandal. 2017. Comprehending Technology Attachment in the Case of Smart Phone-Applications: An Empirical Study. *Journal of Electronic Commerce in Organizations (JECO)* 15 (1): 23–43.

Runswick-Cole, K. 2016. Understanding This Thing Called Autism. In *Re-thinking Autism: Diagnosis, Identity and Equality*, ed. K. Runswick-Cole, R. Mallett, and S. Timimi, 19–29. London/Philadelphia: Jessica Kingsley Publishers.

Runswick-Cole, K., R. Mallett, and S. Timimi. 2016. Introduction. In *Re-thinking Autism: Diagnosis, Identity and Equality*, ed. K. Runswick-Cole, R. Mallett, and S. Timimi, 110–132. London/Philadelphia: Jessica Kingsley Publishers.

Russell, G. 2016. Early Diagnosis of Autism: Is Earlier Always Better? In *Re-thinking Autism: Diagnosis, Identity and Equality*, ed. K. Runswick-Cole, R. Mallett, and S. Timimi. London/Philadelphia: Jessica Kingsley Publishers.

Rutter, Michael. 1972. *Maternal Deprivation Reassessed*. Harmondsworth: Penguin.

———. 2011. Research Review: Child Psychiatric Diagnosis and Classification: Concepts, Findings, Challenges and Potential. *Journal of Child Psychology and Psychiatry* 52 (6): 647–660.

Sarkadi, Anna, Robert Kristiansson, Frank Oberklaid, and Sven Bremberg. 2008. Fathers' Involvement and Children's Developmental Outcomes: A Systematic Review of Longitudinal Studies. *Acta Paediatrica* 97 (2): 153–158.

Sayers, Janet. 1992. *Mothering Psychoanalysis: Helene Deutsch, Karen Horney, Anna Freud and Melanie Klein*. Harmondsworth: Penguin.

Scassellati, Brian. 1993. Implementing Models of Autism with a Humanoid Robot. Yale University, USA. Online: http://www.cs.yale.edu/homes/scaz/abstracts/1999/scaz3.pdf

———. 2002. Theory of Mind for a Humanoid Robot. *Autonomous Robots* 12 (1): 13–24.

———. 2007. How Social Robots Will Help Us to Diagnose, Treat, and Understand Autism. In *Robotics Research*, 552–563. New York: Springer.

Scassellati, Brian, Henny Admoni, and Maja Mataric. 2012. Robots for Use in Autism Research. *Annual Review of Biomedical Engineering* 14: 275–294.

Sevenhuijsen, Selma. 1998. *Citizenship and the Ethics of Care: Feminist Considerations on Justice, Morality, and Politics*. London/New York: Routledge.

Shattuck, Roger, and Douglas Keith Candland. 1994. *The Forbidden Experiment: The Story of the Wild Boy of Aveyron*. New York: Kodansha International.

Shorter, E., 1975. The making of the modern family (pp. 29–30). New York: Basic Books.

Silverman, Chloe. 2011. *Understanding Autism: Parents, Doctors, and the History of a Disorder*. Princeton: Princeton University Press.

Simon, G. 2016. Thinking Systems: 'Mind' as Relational Activity. In *Re-thinking Autism: Diagnosis, Identity and Equality*, ed. K. Runswick-Cole, R. Mallett, and S. Timimi. London/Philadelphia: Jessica Kingsley Publishers.

Simut, Ramona, Cristina Pop, Jelle Saldien, Alina Rusu, Sebastian Pintea, Johan Vanderfaeillie, Daniel David, and Bram Vanderborght. 2012. Is the Social Robot Probo an Added Value for Social Story Intervention for Children with Autism Spectrum Disorders? *ACM*, New York, NY, USA.

Sinclair, Thomas Alan. 1962. *Aristotle the Politics*. Harmondsworth: Penguin Books.

Solomon, Olgo. 2012. Doing, Being and Becoming: The Sociality of Children with Autism in Activities with Therapy Dogs and Other People. *The Cambridge Journal of Anthropology* 30 (1): 109–126.

Sprehe, J. Timothy. 1961. Feral Man and the Social Animal. *The American Catholic Sociological Review* 22 (2): 161–167.

Stawarska, Beata. 2009. *Between You and I: Dialogical Phenomenology*. Athens: Ohio University Press.

Stone, Lawrence. 1977. *The Family Sex and Marriage: In England 1500–1800*. New York: Harper & Row.

Strachey, James, and Sigmund Freud. 1961. *Beyond the Pleasure Principle*. New York: Norton.

Strathern, Marilyn. 1988. *The Gender of the Gift: Problems with Women and Problems with Society in Melanesia*. Vol. 6. Berkeley/London: University of California Press.

Symington, Neville. 1993. *Narcissism: A New Theory*. London: Karnac Books.

Tapus, Adriana, Andreea Peca, Amir Aly, Cristina Pop, Lavinia Jisa, Sebastian Pintea, Alina S. Rusu, and Daniel O. David. 2012. Children with Autism Social Engagement in Interaction with Nao, an Imitative Robot – A Series of Single Case Experiments. *Interaction Studies* 13 (3): 315–347.

Taussig, Michael T. 1993. *Mimesis and Alterity: A Particular History of the Senses*. New York/London: Routledge.

Thill, Serge, Cristina A. Pop, Tony Belpaeme, Tom Ziemke, and Bram Vanderborght. 2012. Robot-Assisted Therapy for Autism Spectrum Disorders with (Partially) Autonomous Control: Challenges and Outlook. *Paladyn, Journal of Behavioral Robotics* 3 (4): 209–217.

Thomson, Mathew. 1998. *The Problem of Mental Deficiency: Eugenics, Democracy, and Social Policy in Britain C. 1870–1959*. Oxford: Oxford University Press.

Tilton, James R., and Donald R. Ottinger. 1964. Comparison of the Toy Play Behavior of Autistic, Retarded, and Normal Children. *Psychological Reports* 15 (3): 967–975.

Timimi, S., and B. McCabe. 2016a. What Have We Learned from the Science of Autism? In *Re-thinking Autism: Diagnosis, Identity and Equality*, ed. K. Runswick-Cole, R. Mallett, and S. Timimi. London/Philadelphia: Jessica Kingsley Publishers.

———. 2016b. Autism Screening and Diagnostic Tools. In *Re-thinking Autism: Diagnosis, Identity and Equality*, ed. K. Runswick-Cole, R. Mallett, and S. Timimi. London/Philadelphia: Jessica Kingsley Publishers.

Trevarthen, Colwyn, Kenneth Aitken, Despina Papoudi, and Jacqueline Robarts. 1998. *Children with Autism: Diagnosis and Intervention to Meet Their Needs*. London: Kingsley.

Trimingham, Melissa. 2010. Objects in Transition: The Puppet and the Autistic Child. *Journal of Applied Arts & Health* 1 (3): 251–265.

Turkle, Sherry. 2012. *Alone Together: Why We Expect More from Technology and Less from Each Other*. New York: Basic Books.

Vlachos, Pavlos A., Aristeidis Theotokis, Katerina Pramatari, and Adam Vrechopoulos. 2010. Consumer-Retailer Emotional Attachment: Some Antecedents and the Moderating Role of Attachment Anxiety. *European Journal of Marketing* 44 (9/10): 1478–1499.

Warwick, Kevin. 2004. *I, Cyborg*. Urbana: University of Illinois Press.

Winnicott, D.W. 1965. *The Family and Individual Development*. London: Tavistock.

Wood, Gaby. 2002. *Living Dolls: A Magical History of the Quest for Mechanical Life*. London: Faber & Faber.

Zingg, Robert M. 1940. Feral Man and Extreme Cases of Isolation. *The American Journal of Psychology* 53 (4): 487–517.

CHAPTER 2

Autism, Social Attachment and Things

2.1 Autism Sociality

Social interaction is not only important for communicating with people around us, but also for a wide range of conceptual capacities. Studies of persons with autism spectrum disorders (ASD, a heterogeneous category involving many persons with social interaction difficulties) show that being social contributes more generally to overall cognitive development. For example, the capacity for imagination, role-play, theory of mind (the ability to imagine thoughts, feelings and agency in other persons, animals and things) and metaphorical understanding is, according to some, severely restricted in autistic persons, as is the ability to anthropomorphise objects, events and situations. This, in my view, shows how sociality is related to a complex range of human abilities. Sociality then, discussed here as interpersonal relationships, impacts on all human beings abilties to develop cognitive and emotional abilties. The question is, do persons with autism have different, rather than absent, social experiences? Can people with autism really exist as nonsocial beings who live in the world as 'systemitising' rather than, shaped, like other humans through interpersonal sociality? Are some psychological experts ignoring the social in autism? While autism as a condition can teach anthropology much about what it means to be social, so too can robots. Research conducted by roboticists in the UK, US and Europe suggest that the presence of humanoid robots seems to improve the social skills of persons with autism.

K. Richardson, *Challenging Sociality*, Social and Cultural Studies of Robots and AI, https://doi.org/10.1007/978-3-319-74754-5_2

I am interested, then, in why a machine that takes on the appearance of a human seems to be more effective than real humans in encouraging social skills in autistic children.

I believe studies of autism and robots may shed new light on these theoretical meanings of the social and interpersonal relationality. Such studies show that human relationships appear to be crucially important and fundamental for a whole gamut of capacities, the most crucial of which is imagining that something can become or stand in place of another thing that it is not, could it be that the ability to take different perspectives rests in interpersonal sociality, because the I exist because of others, and is continually encountering others?

If an I always simultaneously invokes a *you,* what kind of you is the other regarded to be? Immediately after birth, infants activate an automatic social architecture that they use to interact with people around them. Within the first few weeks, they will interact with their carers, through smiling, eye contact, crying, frowning and many other involuntary facial movements. Although this interaction is basic, their carers are usually skilled (or believe they are skilled) in interpreting it. This basic social system develops as the child grows, and this system is supported and assisted by the child's carers who interact with him or her. According to experts, some behavioural signs of autism in children may include behaving placidly, or children that do not sit and 'chat'. Later signs may include displaying repetitive behaviours. Around half of autistic children will develop language, some becoming exceptionally articulate. Language is sometimes developed in a different way among autistic children from children with typical development. For example, children with autism are more likely to use language literally with restricted use of jokes and metaphors (Lyons and Fitzgerald 2004).

There is no way to cure autism, and many experts, parents and persons with the condition refuse to see the condition as an illness, but rather as a different way of being (Bagatell 2010). Cures for autism have become increasingly popular in recent years, however, with diet and parenting styles being seen as the cause of autism. One theory, developed in the 1950s by Kanner and other researchers, proposed that autism developed as a byproduct of 'refrigerator mothering', essentially suggesting that cold and emotionally detached mothers caused autism to develop in their children (Bettelheim 1967). However, this theory has since fallen out of favour.

According to autism literature, autistic persons are severely restrained in their abilities to interact with people, preferring things that demonstrate routine and predictable behaviours (Baron-Cohen et al. 2009). This affects the capacities of autistic persons to interact with other people and restricts their enjoyment of such interactions. More crucially, a lack of sociality appears to contribute to a whole distortion and difference of conceptual and cognitive abilities, such as the ability to imagine, anthropomorphise objects (such as imagine that objects can have intentions or states), mind-read (attribute thoughts and feelings to people), and cope with unpredictability (Baron-Cohen 1997). The term 'autism' is a compound of two Greek words—'aut-', which means 'self', and '-ism', which implies 'orientation or state'. So, autism is a condition where the person is absorbed with his- or her-self (Trevarthen et al. 1998).

Autism was first described clinically in a paper by Leo Kanner in the early 1940s. However, there were earlier descriptions of autistic behaviours, such as 'the boy of Aveyron' whom Itard studied in 1801, and Kaspar Hauser who was 'discovered' in 1828. The accounts of these individuals display the symptoms that Kanner later described (ibid, p. 4). In 1913, a paper by Eugen Bleuler first used the term autism to describe schizophrenics as showing autistic behaviours. For Bleuler, the confusion between reality and fantasy is quite normal when people are being imaginative and creative; only if the balance between autism and realism is lost does a 'pathological behaviour emerge' (4). Kanner was the first to apply this term to a disorder of 'affective contact', and he categorised the condition 'early infantile autism'. According to Kanner's original paper (1943) the main characteristics of an autistic child are:

An inability to establish social relatedness,
A failure to use language normally for the purpose of communication,
An obsessive desire for the maintenance of sameness,
A fascination for objects, and
Good cognitive potentialities.

These characteristics above appear in the child before the age of 30 months.

Interestingly, in 1956 Kanner amended these features to only two:

Aloneness, and
Obsessive desire for sameness

Despite its clinical origins in the 1940s, autism was not included as a diagnostic category in the American Psychiatric Association's Diagnostic and Statistical Manual of Mental Disorders (DSM) until its third edition (DSM-III), published in 1980 (Blacher and Christensen 2011) By the fourth edition of the DSM (1994, pp. 143–147), the inflexibility of the autistic mind was made explicit in descriptions of autism and referred to as 'stereotyped and repetitive motor mannerisms' (Loh et al. 2007, pp. 25–36). In the most recent DSM-5 (American Psychiatric Association 2013), autism spectrum disorder is categorised as one of the neurodevelopmental disorders, an umbrella term which includes attention-deficit/hyperactivity disorder (ADHD) and communication disorders including speech sound disorder and social (pragmatic) communication disorder (2013, p. 31). The manual goes on to outline the main characteristics of autism spectrum disorder:

> Autism spectrum disorder is characterized by persistent deficits in social communication and social interaction across multiple contexts, including deficits in social reciprocity, nonverbal communicative behaviors used for social interaction, and skills in developing, maintaining, and understanding relationships. In addition to the social communication deficits, the diagnosis of autism spectrum disorder requires the presence of restricted, repetitive patterns of behavior, interests, or activities. (ibid)

Timimi and McCabe (2016) note that Marcia Angell, a former editor of the New England Journal of Medicine, criticized DMS-5's open categorisation and said it was one of three 'false epidemics' created by the former edition of the DSM-IV manual (cited in ibid, p. 30). Many biomedical autism psychiatrists and experimental psychologists point to the biological, neurological, and genetic basis to autism etiology (Rutter 2011). However, social and cultural perspectives of medicine, disabilities and difference emphasise that autism is not a stable 'fact' or 'thing' that exists across space and time, pointing to its socially constructed bio-medical origins in the twentieth century (Grinker 2008, 2010, pp. 172–178; Ochs et al. 2004, pp. 147–183; Ochs and Solomon 2010, pp. 69–92). The DSM descriptions might appear neutral and descriptive, but they have prompted analogies between autism and the machine by emphasising the presence of repetitive patterns and obsessive behaviours. The connection with machines is made quite explicit by some: for example, leading autism psychologist, Uta Frith, writes, 'The image I prefer [of the mind] is that of a wondrously complex machine. There are some simple but helpful hypotheses about the work of the mental machinery. For a start we assume that the mind is made up of components which are innately programmed to process information, to produce knowledge and abilities, thoughts and feelings' (Frith 1989).

Baron-Cohen et al. (1997) proposes that autism is an outcome of the Extreme Male Brain (EMB). For Baron-Cohen, the focus on detail explains why science and technology are fields popular with people on the autistic spectrum. This view of researchers in Science, Technology, Engineering and Mathematics (STEM) as 'autistic' is promoted by experts in the field of autism who also claim similarities between the traits of someone with autism and those working in STEM subjects (Baron-Cohen and Hammer 1997, pp. 193–218). Whilst attention and formal intelligence indicators are impaired in most autistic children, there are cases where individuals with autism display exceptional talents in mathematics, music or art (Happé and Frith 2010).

Autism is sometimes presented as a disorder in thinking about thinking, and as such has been viewed by Uta Frith (at UCL) and Simon Baron-Cohen as a form of *mindblindness*, an incapacity to read other people's minds. In his book, *Mindblindness*, psychologist Baron-Cohen uses the term mindreader to describe typical mental states, and coined the term mindblind to refer to persons who cannot recognise or imagine the intentions and states of others. He explains:

> By the end of the first year of life, normal infants…can tell that they and someone else are attending to the same thing, and can read people's actions as directed at goals and as driven by desires. As toddlers, they can pretend and understand pretence. And by the time they begin school, around age 4, they can work out what people might know, think, and believe. According to the model I have outlined, this is due in part to the maturing of…mechanisms that the infant has pre-wired into its brain – its inheritance from a long evolutionary history. (Baron-Cohen 1997)

'The gulf between mindreaders and the mindblind must be vast' writes Gopnik (1993) who gives a vivid account of her attempt to imagine, just what the world must look like through the eyes of someone who is mindblind:

> This what it's like to sit round the dinner table. At the top of my field of vision is a blurry edge of nose, in front are waving hands…Around me bags of skin are draped over chairs, and stuffed into pieces of cloth, they shift and protrude in unexpected ways…Two dark spots near the top of them swivel restlessly back and forth. A hole beneath the spots fills with food and from it comes a stream of noises. Imagine that the noisy skin-bags suddenly moved toward you, and their noises grew loud, and you had no idea why, no way of explaining them or predicting what they would do next. (Gopnik cited in Baron-Cohen 1997, p. 4–5)

Baron-Cohen builds on Daniel Dennett's idea of 'intentional states' which means 'coming up with an explanation of the complex system's behavior and predicting what it will do next' (1995, p. 21).The term 'Intentional Stance' refers to the ability to attribute intentional states (beliefs, desires, thoughts, intentions, hopes, memories, fears, promises, etc.,) to others (Baron-Cohen 1997). Dennett also proposed two alternative stances:

The Physical Stance (understanding systems in terms of their physical make-up)
Design Stance (understanding systems in terms of their functional design).

Dennett is not committed to Intentional States being an outcome of things such as mental states inside the heads of organisms. According to Baron-Cohen, this may be just because Dennett thinks that we consider organisms as rational agents (Baron-Cohen 1997, p. 24). As a thermostat's behaviour is given as an example of an intentional stance, these stances could be applied to entities besides living organisms, such as machines.

What autism is, and who it affects, is not fixed across time and space, and the history of autism is presented differently depending on how it is understood and whose perspective and authority on the subject is prioritised. For example, since 1992, The World Health Organisation (WHO) International Classification of Diseases has been widely adopted outside the US and distinguishes between many types of social-interaction disorders. It breaks up the classification of autism into types, including infantile autism (onset before 30 months), childhood autism (present before 3 years of age), atypical autism (onset after 3 years of age) and other types of related disorders:

Rett's syndrome
Other childhood disintegrative disorder
Overactive disorder associated with mental retardation and stereotyped movements
Asperger's syndrome
Other pervasive developmental disorders
Pervasive developmental disorders

Asperger's syndrome is perhaps one of the most widely known on the autism spectrum disorder scale. It is considered a high-functioning form of autism. The clinical category "Asperger's" follows in a scientific tradition of being named after its founder the Austrian pediatrician Hans Asperger. Asperger published his paper in 1944, one year after Leo Kanner, but it did not receive significant attention until the early 1980s (Frith 1991). According to Timimi and McCabe (2016): 'In the 1970s psychiatrist Lorna Wing saw a similarity in some people she was seeing and those described by Asperger. Dr Wing's ideas intersected with another psychiatrist, Michael Rutter, and formed the basis for the expansion of the concept of autism into Autism Spectrum Disorders' (p. 32). And so Autism Spectrum Disorders promoted an 'expansion' of the concept of autism by Wing and Rutter, which was 'not the result of any new scientific discoveries, but rather new ideologies' (ibid). Persons with Asperger's are able to perform on Model of Mind tests, though they do display a social impairment. Although my study focuses on autism, Asperger's is interesting for a number of reasons, including its increasing diagnosis amongst people in the general population. In recent years there has been an explosion in the number of people diagnosed with autism spectrum conditions. Grinker (2008) noted the increase in diagnoses was in part impacted by the more expansive definition of autism that was included in the DSM-IV, which matches the WHO classifications:

2.2 299.00 Autistic Disorder

An autism screening tool must meet all three primary areas defined by the DSM-IV description for autistic disorder (#'s 1–3 under A below) to qualify for a positive rating from First Signs:

A. A total of six (or more) items from (1), (2), and (3), with at least two from (1), and one each from (2) and (3):

(1) Qualitative impairment in social interaction, as manifested by at least two of the following:

(a) marked impairment in the use of multiple nonverbal behaviors, such as eye-to-eye gaze, facial expression, body postures, and gestures to regulate social interaction
(b) failure to develop peer relationships appropriate to developmental level
(c) a lack of spontaneous seeking to share enjoyment, interests, or achievements with other people (e.g., by a lack of showing, bringing, or pointing out objects of interest)
(d) lack of social or emotional reciprocity

(2) qualitative impairments in communication, as manifested by at least one of the following:

(a) delay in, or total lack of, the development of spoken language (not accompanied by an attempt to compensate through alternative modes of communication such as gesture or mime)
(b) in individuals with adequate speech, marked impairment in the ability to initiate or sustain a conversation with others
(c) stereotyped and repetitive use of language or idiosyncratic language
(d) lack of varied, spontaneous make-believe play or social imitative play appropriate to developmental level

(3) restricted, repetitive, and stereotyped patterns of behavior, interests, and activities as manifested by at least one of the following:

(a) encompassing preoccupation with one or more stereotyped and restricted patterns of interest that is abnormal either in intensity or focus
(b) apparently inflexible adherence to specific, nonfunctional routines or rituals
(c) stereotyped and repetitive motor mannerisms (e.g., hand or finger flapping or twisting or complex whole-body movements)
(d) persistent preoccupation with parts of objects

B. Delays or abnormal functioning in at least one of the following areas, with onset prior to age 3 years: (1) social interaction, (2) language as used in social communication, or (3) symbolic or imaginative play.
C. The disturbance is not better accounted for by Rett's disorder or childhood disintegrative disorder. (DSM-IV 1994)

The National Institute for Clinical Excellence's publication on autism explained the following:

> The prevalence in children and young people of all disorders in the autism spectrum (which includes autism, Asperger's syndrome and atypical autism) has risen in the past decade. The prevalence for all autism spectrum disorders (ASDs) ranges from 60 per 10,000 to more than 100 per 10,000 in the UK. The prevalence for autism is reported to range from 20 to 40 per 10,000.

There is variation in medical diagnoses of autism, but like many contemporary diagnostic tools, it is unclear whether this is due to an actual increase in a particular phenomenon, the introduction of better detective measures, or just the expansion or invention of a category of a particular disorder. *The Essential Difference* (Baron-Cohen 2004) makes a case for a genetic and naturalistic basis to autism linked to the male hormones (as the frequency of males diagnosed with the condition is considerably higher than women). Baron-Cohen divided the male and female brain into the S-brain (for systemising) and the E-brain (for empathising). To illustrate this, Baron-Cohen described a case where someone with Asperger's draws an analogy between people and faeces. The person explains that they have no responsibility for another persons' feelings, because when he has communicated something to others, it is not his fault how people interpret such words, and like faeces, once they have left his body they are no longer his problem.

2.3 Autism and Robots

This picture of human sociality is complicated further by the introduction of humanoid robots. Studies have shown that the presence of a humanoid robot can positively encourage the social skills of children with autism (Breazeal 2009; Dautenhahn and Werry 2004; Peca 2016; Pop et al. 2014; Robins et al. 2005; Scassellati et al. 2012; Thill et al. 2012). This intermediary object, which is neither considered human, nor a mere machine, seems to provide some missing link that is absent in either persons or things.

Across research labs and companies around the world there are many different kinds of experimental and commercially produced robots for children with autism. Two examples are the robot KASPAR (a commonly used robotic platform for autism trials in the UK) and the French manufactured NAO robot (which also has a significant role to play in autism studies). In some cases, 'pretend' robots in the form of humans acting as robots covered in metallic paint have been used. In trials conducted on children with autism, it was found that after interacting with mechanical robots and people taking on the mechanical appearance and behaviours of robots, autistic children displayed more engaged social skills (such as eye-contact, interacting with adults and turn-taking). According to the researchers, there was something about the machine that made these young persons with autism start to behave differently.

Some background on humanoid robotics is required here. It has only really been in the last decade that researchers have started to think seriously about the therapeutic implications of robots for autism. This may have been spurred by a general interest in the technology: the development of humanoid robots began from the mid-1990s in the US and Japan, and now there are dozens and dozens of labs in the world developing socially inspired robots.

I will discuss now some of the early experiments in autism and robotics. One of the first uses of a robot with autistic children came in the 1970s with a robot turtle (Robins et al. 2006). This is how the team in Hertfordshire summarised the work in this field:

> Different from human beings, interactions with robots can provide simplified, safe, predictable and reliable environments that can provide the starting points for therapeutic intervention where the complexity of the therapeutic toys can be slowly increased. (Robins et al. 2005, pp. 105–120)

In one experimental trial, a doll robot called Robota was used. Robota stands about 17 inches tall and had four appearances which the roboticists manipulated, starting from lifelike to mechanical at the end.

State 1 was as a doll
State 2 with its mechanical parts on show (this state was not use in the experimental interactions)
State 3 covered over its body with metallic coloured cloth.
State 4 covered head to toe in metallic coloured cloth.

Robota did seem to encourage social interaction whilst in the experimental setting, but the roboticists could not say if this had any effect outside of the experimental setting. This is important to note. The therapeutic process is understood to be effective if children can generalise and take the experience of the therapeutic encounter into other areas of their lives. As a researcher told me: 'we don't want the children to interact better with robots, but use the experience of interacting with the robot to interact better with people' (fieldwork communication).

2.4 Theatre and Robots

The robot is a creation of theatre, the 'Robot' was a character in the play R.U.R. (Rossum's Universal Robots) in the 1920s, and today the role of theatre and screen fictions shape and direct how the public, and roboticists, study robots. The following is an example of this where theatrical performance was enrolled into a robotics project, to help roboticists learn more about the ways that robots could help children with autism. In a study by researchers at the University of Hertfordshire researchers enrolled a mime artist into robotic studies. One of the lead researchers had a background in therapeutic dance and whilst walking in Covent Garden in London saw an artist performing there, the mime artist was standing like a statue. The researcher thought it would be interesting to see if a human that had a machine-like appearance could provoke the same interest from the children with autism as the Robota robot.

In the first set of tests, the actor was dressed in a suit and was asked to move slowly, lifting one arm at a time, and then lowering the arm, lifting one leg at a time, then alternating with the other leg. What the researcher told me next was unexpected.

> We had the actor dress in a suit and move slowly and mechanically. The child did not interact with the actor and just tended to ignore him.

Then the same actor was dressed head to toe in a metallic fabric and his face was also covered. The actor moved in exactly the same way as before, but this time the child ran to the actor and grabbed his leg. I saw still images of these two different scenarios. In one still image there was little interest in the human-looking actor by the children with autism, but in the still with the theatrical robot, the child was in closer proximity, touching the actor's arm. These images were still images and images capture information in a particular kind of way. There is no way of knowing what happens before, during or after each event, nor why the children became fascinated with a mechanical-looking human. What perhaps is more interesting about the rationale for the robot is provided in a paper by these researchers, who wanted to test if a robot's appearance impacted how autistic children interacted with it. When it came to using a full-sized humanoid robot, there was neither the financial nor technical resources to produce such a machine (humanoid robots are extremely complex and expensive platforms). They explain:

> In order to address this lack of the availability of an easy to control, safe, life-sized humanoid robot that can be used in the studies, we developed a novel approach by using a Theatrical Robot – a mime artist – a person who was dressed and acted like a robot. The Theatrical Robot is a life-sized, embodied, 'simulated robot' which allows to investigate requirements of robot design even prior to any hardware and software development. The Theatrical Robot consists of a human instructed to behave and/or appear like a robot. (Robins et al. 2006, p. 485)

There are no hard and fast rules that say a robot is grey, or made of metal-*like* material. Thus the experiment did not recreate a robot per se but a particular kind of portrayal of a robot. Robots can be made of plastic, and/or multicoloured. The NAO[1] robot, produced by Aldebaran robotics is red and white or blue and white. The Pepper[2] robot created by Softbank/Aldebaran robotics is white. The robot ASIMO,[3] produced by the Honda corporation, has a white body-frame and a black face mask. The Keepon[4] robot, is small, yellow and fits on a table-top. The Probo[5] robot, developed by researchers at *Vrije Universiteit Brussel,* is green and looks like a cross between an elephant and a fantasy figure. The authors of the study took their idea of a robot and projected it onto a human actor, who was then described as being a robot. The authors of the paper might have been drawing their idea of the robot from fictions written between the late 1920s to the 1950s, when the association between robots and metal contraptions became a dominant form. The representation of a robot as a machine, or made of metal, was not even part of the earliest representations of robots.

[1] NAO robot. Accessed 20.8.17. https://www.ald.softbankrobotics.com/en/robots/nao

[2] Pepper robot. Accessed 20.8.17. https://www.ald.softbankrobotics.com/en/robots/pepper

[3] ASIMO robot. Accessed 20.8.17. http://asimo.honda.com/

[4] Keepon robot. Accessed 20.8.17. http://www.mykeepon.com/buy

[5] Probo robot. Accessed 20.8.17. http://probo.vub.ac.be/Probo/

The characters of Robots[6] in the play *R.U.R.*, written by Karel Čapek in 1920 and first performed in 1921 in Prague, were humanlike in appearance to the human characters in the play. The robots in *R.U.R* looked identical to the non-robotic humans in the play; the only difference was their lack of will, subjectivity and inner experience. They were created as tools (slaves) to be used by managers and industrialists to work; the term robot comes from the Slavic term robota: 'forced labour' (OED 1995, p. 1191). It was not the appearance of the robots that made them different from humans, but what was inside of them, that could not be seen on the surface. It was other artists in the 1920s and 1930s, along with engineers who turned robots into machines, and made a metallic like appearance synonymous with robots. Fritz Lang, the avant-garde German director, and his artistic team, introduced one of the earliest fictional representations of a metallic human, but the term 'robot' was not used at all in the film; instead the term 'Man-Machine' was used to describe the metallic Maria in the film *Metropolis* (1927).

By the 1930s, the robot had become synonymous with a machine form, often grey and metallic. Even Čapek was concerned about the way in which other artists had interpreted the robot to mean something other than he intended. Čapek had wanted to use the Robot to talk about the obsession with work and the reducibility of human beings to labour in capitalist and communist narratives of the 1920s.

It is not clear what the children in the study considered to be a 'robot'. In the contemporary era robots come in many shapes and sizes, fictionalised in films such as Star Wars (from 1977–2005 and beyond), or in cartoons, games and toys. Without knowing what conception the children had of robots, there is no way of deducing that the children were responding to the mime artist *as a* robot. The children in the study could have been responding to the unsual looking mime artist behaving strangely in a corner of the room. Here I draw attention to the assumptions made by researchers of their own idioms as the starting point for others. This is a case of what Michael Taussig explores in his study of colonial encounters with indigenous peoples in his book, *Mimesis and Alterity: A Particular History of the Senses* Taussig's extrordinary book explores how colonialists use something intrinsic to their own culture as a reference point, and *project it onto* other peoples' as if it were *a shared* reference point (1993).

[6] Karel Čapek, author of the play, Rossum's Universal Robots (R.U.R) always capitalized the I.

I have discussed elsewhere in my work how roboticists often act in place of their machines (Richardson 2015). Whilst conducting my graduate fieldwork at the Massachusetts Institute of Technology (MIT) robot labs, I would attend robot demonstrations given to members of the lab, the public or potential sponsors. Often these machines would break down, either due to a flaw in the code or some other technical glitch. If the robot failed to perform in its actions, the roboticist would stand in its place and demonstrate the movement, slowing down their motion to the speed of the robotic machine and breaking up the bodily movements into discrete parts. A hand opening a door is a grasp, a twist, a push. I write about this elsewhere as a practice of mimesis—and find that robots are always tied to the minds and the bodies of people who create them (Richardson 2010, pp. 75–90).

In this particular case, the humanoid robot was too costly and complicated, so a human being took on the role of the imagined machine. The group concluded that the more machine-like the appearance and behaviours of the doll and the human actor, the more the children with autism would interact with it. The more human-like, the more likely they were to ignore the human actor and robot. I asked the researcher, Josef, why he thought this was so. He was not sure but speculated that it might be to do with the with the simplicity of the machine and the person when they are covered up. This prompts the question: to elicit interaction from the children with autism, did it have to be grey metallic fabric or would any other fabric have done as long as the doll or person looked less human-like?

The argument that the more machine-like the robot, the more effective it is at inspiring interaction with children with autism is somewhat challenged by the introduction of the robot Gizbar. Gizbar, at the time of this study, did not have any innovative technology. Gizbar sits at less than 1 metre tall, its lower body is placid and immobile, whilst its upper body, arms, and head are movable. On his face, his eyes have two degrees of freedom, which means they can move in two directions, the eyelids move, and facial movements resembling smiles or interest can be mimicked by the machine. Gizbar wears clothes, and although it is meant to be genderless many people perceive Gizbar as a boy. In fact, during my first days in the lab I was told by a female researcher that she tried to 'turn Gizbar into a girl' by changing its clothes and putting a long blonde wig on the robot, but the professor of the lab (a woman) decided against it, saying Gizbar did not have a gender. Despite efforts to neutralize Gizbar's outward appearance, it does wear boys' clothes and was given a boy's name.

I first met Gizbar at the Manchester Science Festival. I was impressed by its physical appearance: it looked very different from the very mechanical looking machines at MIT, but it did strike me as slightly creepy. In fact, at one of the public events, a teenage boy entered the room and within a few minutes could not resist talking about how scary he thought it was.

The reasons given by roboticists and autism experts why autistic children seem to like robots are as follows:

> robots appear more predictable and less complex than people. Details overwhelm persons with autism. Having a more simplified person might be part of the appeal.

Roboticists believe a route to supporting persons with autism is, in part, to be able to enter their world. But this begs the question: how do they understand the lifeworld of a person with autism? Who has constructed this understanding? Is it best to enter into the autistic person's world through *things* rather than engaging with them interpersonally? Does decorating objects with social qualities help this process? The robot as an intermediary!

Roboticists use robots as an intermediary in this process, but does the robot really function as such or does this claim only appear reasonable because of how roboticists interpret the behaviour of children with autism? Take the following study by researchers at the Autism Research Centre in Cambridge called *The Transporters Project*. The Transporters is a video given to children with autism and their families as a teaching tool to learn about emotions. The makers of The Transporters write:

> Children with autism tend to love vehicles, probably because they are not alive and don't move unpredictably. What they seem to dislike are objects that move when they least expect them to, and for no apparent reason. In the world of vehicles, some forms of transport are more predictable than others. For example, cars and planes can move in almost any direction, at the whim of the driver. In that sense, cars and planes are not much better than people – they are hard to predict. Other kinds of vehicles are much more constrained in their movements. Trams have to go in straight lines because they are attached to tracks and overhead cables. So do cable cars and trains and funiculars. If you think about it, such vehicles can only go forward and backward or along the gentle curve of a track. These are the vehicles most loved by children with autism. (Users Guide, The Transporters 2006, pp. 4–5)

2.5 Anthropology and the Social

Anthropology has long had an interest in exploring the limits of modernist meanings of the 'social' and so anthropological theorising has contributed to extending the social to non-human animals and things. The social can be embedded in things which can stand in for persons such as gifts or art objects (Strathern 1988; Gell 1998). Latour (2005, 2012) suggests the social is not exclusively human and proposes hybrids of persons and things in networks rather than purity of these categories, while Haraway extends the notion of the social to animals, notably primates and dogs (2003). But is anthropology right to extend the social to include nonhumans? Do persons play a role that is greater than or different from that of any other entity?

Rather than downgrade the human role in social relations as these scholars suggest, we might find another way to rethink the relations between persons and things by studying autism and robots. For me, these studies show how human sociality and understanding of others is important for helping human beings to enjoy life, but is also, as I have shown earlier, intimately connected with other capacities. Studies in autism show how fundamentally social-communication impacts other areas of life. In autism for example, a difficulty in social interaction impacts on the metaphorical imagination. Of course, children and adults with autism have imagination, but metaphorical imagination concerns the substitution of forms—one thing standing in for others, taking the place of others in a system of symbolic references (Lakoff and Johnson 2008; Lakoff 1987).

In a study published on toys and autism, psychologists Tilton and Ottinger (1964) suggested that what is lacking in autistic persons is not a capacity to understand people, but the capacity to understand the properties of objects. They wrote that the 'inability to perceive relationships between objects might contribute to their failure to develop social skills, relations between child and things, and things and things being conceived in the same way as relations with persons' (p. 51). I want to reverse this position and propose that it is not the inability to perceive relationships between objects, but the difficulties in making sense of human relationships that impacts on the way children with autism relate to objects.

Children with autism do characteristically treat objects differently from how they treat other people. This is not to say they do not relate at all to people, and like all objects the same, but their relations with their physical environment and relations with others have different characteristics. What underlies this difference? Freud asks a question here that I will explore in depth through this monograph. In *Beyond the Pleasure Principle* (1961), whilst lodging in a house, Freud noticed a game that a child used to play with his parents. The child was one and half, with few intelligible words. Freud describes the child as pleasant child who was often praised by one servant-girl as a 'good boy', and as a boy who rarely disturbed his parents, and showed great affection towards his mother. I will now go on to cite the passage:

> This good little boy, however, had an occasional disturbing habit of taking any small objects he could get hold of and throwing them away from him into a corner, under the bed, and so on, so that hunting for his toys and picking them up was often quite a business. As he did this he gave vent to a loud, long-drawn-out 'o-o-o-o, accompanied by an expression of interest and satisfaction. His mother and the writer of the present account were agreed in thinking that this was not a mere interjection but represented the German word 'fort? ['gone']. I eventually realized that it was a game and that the only use he made of any of his toys was to play 'gone' with them. One day I made an observation which confirmed my view. The child had a wooden reel with a piece of string tied round it. It never occurred to him to pull it along the floor behind him, for instance, and play at its being a carriage. What he did was to hold the reel by the string and very skillfully throw it over the edge of his curtained cot, so that it disappeared into it, at the same time uttering his expressive 'o-o-o-o'. He then pulled the reel out of the cot again by the string and hailed its reappearance with a joyful 'da' ['there'].
>
> This, then, was the complete game disappearance and return. As a rule one only witnessed its first act, which was repeated untiringly as a game in itself, though there is no doubt that the greater pleasure was attached to the second act. The interpretation of the game then became obvious. It was related to the child's great cultural achievement the instinctual renunciation (that is, the renunciation of instinctual satisfaction) which he had made in allowing his mother to go away without protesting. He compensated himself for this, as it were, by himself staging the disappearance and return of the objects within his reach. (Freud 1995, p. 599–600)

I conclude this chapter with Freud as a way of departure, of reflecting on relations with the physical world and intimate relationships, before I discuss the models of attachment. This passage by Freud is part of the puzzle of attachment, and how human attachment helps shapes a person's relations with themselves, others and the world around them. Freud is suggesting the primary attachments are to childhood core relationships—the mother in this case, the game only makes sense because of these intimate attachments.

Bibliography

American Psychiatric Association. 1994. DSM-IV: *Diagnostic and Statistical manual.* Washington, DC: American Psychiatric Association.

———. 2013. *Diagnostic and statistical manual of mental disorders (DSM-5®).* Washington, DC: American Psychiatric Publishing.

American Psychiatric Association, and American Psychiatric Association. 1994. *Diagnostic and Statistical Manual of Mental Disorders (DSM)*, 143–147. Washington, DC: American Psychiatric Association.

Asperger, A. 2001 [1944]. 'Autistic Psychopathy' in Childhood. Trans. and annotated by Uta Frith. In *Autism and Asperger Syndrome*, ed. U. Frith. Cambridge: Cambridge University Press.

Bagatell, Nancy. 2010. From Cure to Community: Transforming Notions of Autism. *Ethos* 38 (1): 33–55.

Baron-Cohen, S. 1997. *Mindblindness: An Essay on Autism and Theory of Mind.* Cambridge: MIT Press.

———. 2004. *The Essential Difference.* London: Penguin.

Baron-Cohen, Simon, and Jessica Hammer. 1997. Is Autism an Extreme Form of the "Male Brain"? *Advances in Infancy Research* 11: 193–218.

Baron-Cohen, Simon, Sally Wheelwright, Carol Stott, Patrick Bolton, and Ian Goodyer. 1997. Is There a Link Between Engineering and Autism? *Autism-London* 1: 101–109.

Baron-Cohen, Simon, Ofer Golan, Emma Chapman, and Yael Granader. 2009. Transported to a World of Emotion. *McGill Journal of Medicine* 12 (2): 78–81.

Bettelheim, Bruno. 1967. *Empty Fortress.* New York/London: Free Press/ Collier-Macmillan.

Blacher, Jan, and Lisa Christensen. 2011. Sowing the Seeds of the Autism Field: Leo Kanner (1943). *Intellectual and Developmental Disabilities* 49 (3): 172–191.

Breazeal, Cynthia. 2009. Role of Expressive Behaviour for Robots That Learn from People. *Philosophical Transactions: Biological Sciences* 364 (1535): 3527–3538.

Dautenhahn, Kerstin, and Iain Werry. 2004. Towards Interactive Robots in Autism Therapy: Background, Motivation and Challenges. *Pragmatics & Cognition* 12 (1): 1–35.

Frith, U. 1989. Autism: explaining the enigma. Oxford: Basil Blackwell.

———. 1991. *Autism and Asperger Syndrome.* Cambridge: Cambridge University Press.

Freud, Sigmund. 1995 [1920]. Beyond the Pleasure Principle. In *The Freud Reader*, ed. Peter Gay, 594–626. London: Vintage.

Gell, Alfred. 1998. *Art and Agency: An Anthropological Theory.* Oxford: Clarendon Press.

Gopnik, Alison. 1993. Theories and Illusions. *Behavioral and Brain Sciences* 16 (01): 90–100.

Grinker, Roy Richard. 2008. *Unstrange Minds: Remapping the World of Autism*. New York: Basic Books.

———. 2010. Commentary: On Being Autistic, and Social. *Ethos* 38 (1): 172–178.

Happé, Francesca, and Uta Frith. 2010. *Autism and Talent*. Vol. 364. Oxford: Oxford University Press.

Haraway, Donna Jeanne. 2003. *The Companion Species Manifesto: Dogs, People, and Significant Otherness*. Vol. 1. Chicago: Prickly Paradigm Press.

Kanner, L. 1943. Autistic Disturbances of Affective Contact. *Nervous Child* 2 (3): 217–250.

Lakoff, George. 1987. *Women, Fire, and Dangerous Things. What Categories Reveal About the Mind*. Chicago: The University of Chicago Press.

Lakoff, George, and Mark Johnson. 2008. *Metaphors We Live By*. Chicago: University of Chicago Press.

Latour, Bruno. 2005. *Reassembling the Social: An Introduction to Actor-Network-Theory*. Oxford/New York: Oxford University Press.

———. 2012. *We Have Never Been Modern*. Cambridge, Mass: Harvard University Press.

Loh, Alvin, Teesta Soman, Jessica Brian, Susan E. Bryson, Wendy Roberts, Peter Szatmari, Isabel M. Smith, and Lonnie Zwaigenbaum. 2007. Stereotyped Motor Behaviors Associated with Autism in High-Risk Infants: A Pilot Videotape Analysis of a Sibling Sample. *Journal of Autism and Developmental Disorders* 37 (1): 25–36.

Lyons, V., and M. Fitzgerald. 2004. Humor in Autism and Asperger Syndrome. *Journal of Autism and Developmental Disorders* 34 (5): 521–531.

Metropolis (1927) Dir. Fritz Lang.

Ochs, Elinor, and Olga Solomon. 2010. Autistic Sociality. *Ethos* 38 (1): 69–92.

Ochs, Elinor, Tamar Kremer-Sadlik, Karen Gainer Sirota, and Olga Solomon. 2004. Autism and the Social World: An Anthropological Perspective. *Discourse Studies* 6 (2): 147–183.

Oxford English Dictionary (OED). 1995. *The Concise Oxford Dictionary of Current English*. First edited by H.W. Fowler and F.G. Fowler. Ninth Edition. Edited by Della Thompson. Oxford: Clarendon Press.

Peca, Andreea. 2016. Robot Enhanced Therapy for Children with Autism Disorders: Measuring Ethical Acceptability. *IEEE Technology and Society Magazine* 35 (2): 54–66.

Pop, Cristina A., Sebastian Pintea, Bram Vanderborght, and Daniel O. David. 2014. Enhancing Play Skills, Engagement and Social Skills in a Play Task in ASD Children by Using Robot-Based Interventions. A Pilot Study. *Interaction Studies* 15: 292–320.

Richardson, Kathleen. 2010. Disabling as Mimesis and Alterity: Making Humanoid Robots at the Massachusetts Institute of Technology. *Etnofoor* 22 (1): 75–90.

———. 2015. *An Anthropology of Robots and AI: Annihilation Anxiety and Machines*. Vol. 20. New York: Routledge.

Robins, Ben, R. Te Boekhorst Kerstin Dautenhahn, and Aude Billard. 2005. Robotic Assistants in Therapy and Education of Children with Autism: Can a Small Humanoid Robot Help Encourage Social Interaction Skills? *Universal Access in the Information Society* 4 (2): 105–120.

Robins, Ben, Kerstin Dautenhahn, and Janek Dubowski. 2006. Does Appearance Matter in the Interaction of Children with Autism with a Humanoid Robot? *Interaction Studies* 7 (3): 509–542.

Rutter, Michael. 2011. Research Review: Child Psychiatric Diagnosis and Classification: Concepts, Findings, Challenges and Potential. *Journal of Child Psychology and Psychiatry* 52 (6): 647–660.

Scassellati, Brian, Henny Admoni, and Maja Mataric. 2012. Robots for Use in Autism Research. *Annual Review of Biomedical Engineering* 14: 275–294.

Strachey, James, and Sigmund Freud. 1961. *Beyond the Pleasure Principle*. New York: Norton.

Strathern, Marilyn. 1988. *The Gender of the Gift: Problems with Women and Problems with Society in Melanesia*. Vol. 6. Berkeley: University of California Press.

Taussig, Michael T. 1993. *Mimesis and Alterity: A Particular History of the Senses*. Hove: Psychology Press.

Timimi, S., and B. McCabe. 2016. Autism Screening and Diagnostic Tools. In *Re-thinking Autism: Diagnosis, Identity and Equality*, ed. K. Runswick-Cole, R. Mallett, and S. Timimi, 30–48. London/Philadelphia: Jessica Kingsley Publishers.

The Transporters Discover the World of Emotions, DVD. 2006. Narrated by Stephen Fry. Catalyst Studios.

Thill, Serge, Cristina A. Pop, Tony Belpaeme, Tom Ziemke, and Bram Vanderborght. 2012. Robot-Assisted Therapy for Autism Spectrum Disorders with (Partially) Autonomous Control: Challenges and Outlook. *Paladyn, Journal of Behavioral Robotics* 3 (4): 209–217.

Tilton, James R., and Donald R. Ottinger. 1964. Comparison of the Toy Play Behavior of Autistic, Retarded, and Normal Children. *Psychological Reports* 15 (3): 967–975.

Trevarthen, Colwyn, Kenneth Aitken, Despina Papoudi, and Jacqueline Robarts. 1998. *Children with Autism: Diagnosis and Intervention to Meet Their Needs*. London: Kingsley.

CHAPTER 3

The Experiment: The Effectiveness of a Humanoid Robot for Helping Children Develop Social Skills

3.1 Between the Social and Asocial

A robot occupies an 'imagined' duality, standing between human and machine simultaneously, made up of parts, and assembled in forms that reflect back to us familiar anthropomorphic forms. The child with autism is also characterised as somewhere in-between, standing someway between the social and the asocial. To some experts, such as Baron-Cohen, persons with autism experience an existence free of the trappings of typical sociality—recognition of emotions, reciprocal interaction, joint attention, turn-taking, imitation or social deception. This freedom is what allows them to develop 'hyper-systematising' talents that are special islets of gifts that some children with autism possess to an extraordinary degree (Happé and Frith 2010; Baron-Cohen et al. 2009, pp. 1377–1383).

Autism as a condition does not exist across time and space, but is a category formulated out of normative categories developed in psychiatry and the psychological sciences about what is 'typical sociality'. In the field of robot therapy for children with autism, the robot, in its anthropomorphic form, carries within it ways of being a human without the complexity of actually being one. This reduced resemblance supports the child with autism and can help him or her develop social skills (Robins et al. 2005, pp. 105–120; Pop et al. 2014). Moreover, mechanical analogies with autism are pervasive in the field, from description of behaviours as 'stereotyped and repetitive motor mannerisms' (Loh et al. 2007, pp. 25–36) to descriptions

K. Richardson, *Challenging Sociality*, Social and Cultural Studies of Robots and AI, https://doi.org/10.1007/978-3-319-74754-5_3

of mechanistic psychology underlying the dominant naturalistic models in biomedical psychiatry (Frith 1989). As Danforth and Naraian (2007, pp. 273–290) explain, 'the child or adolescent with autism is being described through an overarching human as machine conceptual metaphor that maps common meanings about machines onto our understanding of flesh and blood person' (p. 273).

It is worth asking, then, why are robots and children with autism put on a par with each other? And does this analogy benefit children with autism or does it come out of a psychological worldview that frames what it means to have autism? As Hacking (2009, pp. 1467–1473) notes, the growth of 'Autistic Biography' means that expert positions are increasingly being challenged. Does becoming like a machine pose any threat as it once did? Are we all cyborgs now? Bruno Bettleheim (1959) believed that mental health disorders and mechanical analogies were synonymous, that moving in the direction of the machine indicated movement away from affective relations. Joey: A 'Mechanical Boy' provides 'a case history of a schizophrenic child who converted himself into a "machine" because he dare not be human. His story sheds light on emotional development in a mechanized society' (p. 4). In a hybrid world where analytical divisions are dissolved, affective relations can include machines (Haraway 2003), our 'silicon others'. Anthropological studies of autism can contribute a richer appreciation to studies of autism than psychiatric descriptions that focus on clinical symptoms. Ochs and Solomon (2010, pp. 69–92) propose an 'autistic sociality':

> Autistic sociality is not an oxymoron but, rather, a systematically observable and widespread phenomenon in everyday life. Without disregarding or underestimating the social impairments related to this condition, we reach toward an account that comprehends both limitations and competencies of autistic sociality. Our ethnographic study of the daily lives of children with autism indicates that their sociality varies across different situational conditions. (p. 69)

My own ethnographic encounters with autism and robots took the form of planning and running a psychology experiment to test the effectiveness of a humanoid robot in encouraging social behaviours in children with autism. Situated within the framework of a psychological experimental process, I conducted a small experimental trial (n = 29 children), and

I embraced the opportunity to be a fully engaged participant observer. In the process, I took on the responsibilities for the experimental design, ethics, assessments, and designing protocols and coding schemes. The experiment was designed by an autism expert and the robot was provided by the University of England. I anonymise all of the persons and only refer to the robot, Gizbar by name. Gizbar is a well-known robot in the field of autism studies. My role in the experiment enables me to actively participate in the subject and see how knowledge is produced.

The opportunity to carry out the experiment in a school for children with autism afforded me well over 200 hours of face-to-face time with children with autism, mainly Low Functioning Autism (LFA), and dozens of hours of video-taped sessions of myself and a child interacting with Gizbar. To complement the novelty and excitement of Gizbar, the team decided to use a contrasting object—a non-humanoid toy crane (the control object in the experiment) and toys.

I also sometimes observed the children playing at break-times while I chatted to the teachers on duty and took the children out for trips to local parks and on activities to museums.

3.2 Robots

There is a growing community of robotics researchers who advocate the use of robots for helping children with autism (Coeckelbergh et al. 2016; Pop et al. 2013, 2014; Tapus et al. 2012; Robins et al. 2006; Bird et al. 2007; Kozima et al. 2005; Simut et al. 2012; Peca 2016; Thill et al. 2012; Robins et al. 2005; Scassellati et al. 2012; Breazeal 2009). The development of robotics for children took place independently of research in the psychological sciences and was spearheaded by researchers in the US and the UK (Scassellati; Dautenhahn and Werry 2004). Robots were classed as 'assistive technologies', new kinds of technologies to provide assistance to children and adults with particular healthcare needs (Feil-Seifer and Mataric 2005).

Why a humanoid robot? A humanoid robot is a mechanical entity that resembles an anthropomorphic form and may move its body in a way intended to mimic human behaviours and expressions. For example, two shoulders lifting simultaneously indicates the feeling 'I don't know'. Putting one's hand to one's chin and tilting the head to the side might mean 'I'm curious' or 'I'm thinking'. As the robot used in our experiment was non-verbal, it was purely ‘expressive’ in its working mode. The body is the medium of communication in robotics, especially humanoid and behaviour-based robotics. The body is never independent of language—it does not exist before, after or separate from language; the body is the framework in which verbal speech and language is expressed. The body is the vehicle for imprinting words on to textures—be it paper or a screen, directly or indirectly using hands or eyes. The body is speech.

The robot Gizbar is an example of a popular robot. Gizbar has immobile legs, but motors in the arms and head, allowing the robot to appear expressive. During the experiment the robot had no autonomy—but its creators believed it to give the ‘impression of autonomy’.

3.3 The School

The school was located in London; a building that was a throwback to the Victorian era, with high foreboding walls and supersized windows that made it almost impossible to control the flood of light in the inner spaces. For children with autism, sensory information is challenging—I often saw the children covering their heads with blankets or jackets, sitting in the corners of these over-stimulated rooms, sheltering themselves from ‘perceptual overload’ (Frith 1991). It existed at least: a school where the specific needs of the children were catered for exclusively; a ‘school for children with autism’; a heterotopia—‘those in which individuals whose behaviour is deviant in relation to the required mean or norm are placed’ (Foucault and Miskowiec 1986, pp. 22–27). Special schools act to contain that which is different, providing a *habitus* of ritual and repetition of social learning, conformity and structured normativity (Bourdieu 1980). The irony is that for children with autism the structure is described as soothing, too much unpredictability being unwelcome.

For admission to the school, children must have received an autism diagnosis from a specialist. Such specialised schools are rare and usually special schools are designed to take a variety of different children from a range of learning and physical difficulties. The teachers, teaching assistants and headteacher were female. The children were related to individually; the teachers and assistants were aware of their mannerisms and behaviours. John, for example, was a student that had very specific behaviours when it came to food: he ate one pea at a time on his plate. The staff supported and accommodated the children as best they could, assisting with these activities patiently. Another child might sit in a corridor with an assistant, rocking back and forth for comfort. Headsets were given to the children to reduce noise and windows covered with makeshift blinds. A physical space for a child with autism is complex, and the stimuli can overwhelm the students. So much of our ability to cope with sensory information is mediated by the senses, which in turn is mediated by relationships. Infants and children cannot self-regulate their emotions: the caregiver soothes a child through social interaction, the child experiences this soothing and can imitate the process. If the social communication is disrupted, the stimuli is overwhelming and the child develops alternative practices to self-soothe. These are referred to in the literature as 'non-functional' because they are not directed towards an outward interaction but are primarily for the child.

In the social sciences, the role of schooling has been central to the creation of normative subjects (Bourdieu 1980), but 'special schools' are often treated differently, as if schooling addressed to children with special needs were outside the frame of discussion of 'cultural capital'. Cultural capital is used in Bourdieu's general theory of capital and habitus to explain 'the unequal scholastic achievement of children originating from the different social classes by relating academic success, i.e., the specific profits which children from the different classes and class fractions can obtain in the academic market, to the distribution of cultural capital between the classes and class fractions' (Bourdieu 2011, pp. 81–93). In representing the school as a socially-inscribed space—the school's structure mirrored typically-developing school structures—classes were organised chronologically rather than by ability. Students were compartmentalised into specific classrooms, organised by year. A syllabus was followed, and the school was subject to national assessment criteria laid out by Ofsted (the Office for Standards in Education, Children's Services and Skills), a non-ministerial department of the UK government, reporting to Parliament via the Department of Education. The school was awarded the highest grade of Outstanding in all areas of assessment in 2014.

At the school, I saw Jonathan, who would hum the same tune, at the end of which he would knock over his chair. He repeated this action over and over again. The teacher tried to encourage him to sit in his chair, but his attention was focused elsewhere.

Soothing behaviours serve an important purpose for children with autism. The school experience has shifted significantly and not too long ago such behaviours would meet with disproval. The physical environment was an important platform for the children to interact with. As I brought Jason up and down the stairs to the robot room, he would lick the interior windows as if engaged with the physical space in a profoundly different way than expected. He would perform this ritual on the way to the lab room, and on the way back down to his classroom.

Schools are rule-bound places, but this school was similar to many others, with classrooms organised around age, not ability, a playground like any other and the day structured around arrival, work, lunch, playtime and home time. It was sometimes strange to meet them only in the moments in the school, I rarely saw the children interacting with their caregivers and never saw any of them in their home environment. The head-teacher was excited about the prospect of a new school for the children, one with a more effectively controlled physical environment and that better conformed to the children's needs. Unfortunately, a new government headed by a new party had not prioritised funding for new school buildings. The project was scrapped. There was a general feeling of depression in the place. I interacted with children from a wide age range, some of them newly arrived infants, or 4 or 5, one or two at the senior end of the school age—13 and 14. Children with autism will grow into adults, and some of these young adults towered over the staff. For the roboticists, robots held the answer for bringing about socially inspired transformations in children with autism, but the reality is one of complex, knotted day to day interactions built up of thousands of interactions and experiences. The school remains, the children attend and the staff work hard, as resources are stretched in new economic circumstances.

3.4 The Protocols

Experiments are carefully constructed practices where specific phenomena can become isolated from other general phenomena to arrive at some results or 'truth' about statements. There are two artefacts created by this process, which are quite different but interconnected. On the one hand there is the experiment, which is geared towards finding something out. On the other hand there is an experimental research paper for describing the experiment. Experimental processes have a similar structure. A hypothesis is created, then an experimental design to test the hypothesis. Then there is the experiment, followed by analysis of results, and findings. The procedure for experiments consists of: participant recruitment, ethical proposals and consent, and hypothesis formulation—what the researchers expect to find.

The experiments in question followed a formulistic pattern that was played out repeatedly in the same way, as if following a script. The experiment was organised as follows:

Method

The study subjects were all children with autism. The ages of our subjects ranged from 5 to 13 years old. Each child received fifteen minutes of interaction with either Gizbar the robot or the toy crane. Each of the sessions was free play with up to 11 sessions per child. A pre- (Time 1) and post-evaluation (Time 2) of the child's social abilities was made through a standardised study designed specifically for the project. Each session was video-taped and coded for the following measures: (1) number of times child made eye-contact with the experimenter, (2) number of times child interacted with experimenter.

Analysis

Data analysis was done in the following ways. Data drawn from the evaluation in pre- (Time 1) and post- (Time 2) sessions were studied for changes in behaviour. Additionally, the codes from the video-taped sessions were studied and combined with data from the evaluations to examine whether the humanoid robot influenced the types of social behaviours shown by the children.

Play Assessment, Session Content and Schedule

The pre-experiment and post-experimental play assessment involved ascertaining the child's abilities in two key areas—eye-contact and turn-taking. Games included hiding toys from the child (under a cloth), blocking toys (so that the child has to initiate contact with the experimenter to retrieve the toy), and turn-taking with the child (which will be initiated with the use of blow-up-bubbles). The three main techniques used were:

- Blocking – The child was "blocked" from engaging with a toy
- Out of Reach – The toy was placed "out of reach" so that the child had to initiate contact
- Teasing – The researcher blew bubbles or blew-up and deflated a balloon to encourage eye contact.

These play activities were performed at the beginning and at the end of the experiment.

During the experimental sessions, play with each individual child focused on exploring the child's behaviours in the following: (1) eye-contact and (2) turn-taking. Play exploring eye-contact included peekaboo, eye-pointing games (where the child has to guess where the robot is looking). Play was encouraged that explored turn-taking by allowing the child to interact with the robot and mimic the robot's actions. The robot behaved in a pre-programmed way and was only able to move via a remote-control (which was held by the researcher or child). The robot had no lower-body movement and limited upper-body movement, though able to move its arms up and down and bend at the elbows for playing games such as peekaboo. The robot could smile and frown and its eyelids could open and close. The robot had no mobility in the hands, but was able to use its hands to bang a drum (a game the children were encouraged to imitate).

For the control object we used a brightly coloured, remote-controlled crane. The crane was nonhumanoid but carried with it certain features shared by the robot, including its colouring, the fact that it was remote controlled and that parts of the crane could be removed (blocks). I encouraged the child to play games via the remote-control (just as with the robot).

3.5 The Experiment

As the protocol described above, each procedure was scripted in advance. I was first introduced to the children individually as they were brought to the nurse's room—the setting for the experiment for the next few weeks. Around the room were two cameras, and laid on the floor were toys of various kinds—trains, books, dolls, cars, tracks. The children entered the room and immediately began to interact with the toys. I would sit with them on the floor, not interacting until the child engaged with me. I was struck on occasion by the ease with which they settled into the play. Other children entered the room and were more interested in the computer on the desk, which recurrently ran an image of people of different ages with vivid facial expressions. Our research team wondered if that might encourage engagement during the pre- and post- play protocol. It was clear that interacting with the toys and the computer was frequently preferred over interacting with me. Some children played with the objects in typical ways but one student, Sarah, took each toy, and built them up into a tall totem.

All the children whose parents consented to them taking part in the experiment did so. The group was divided between those children presented with the robot, and those given the crane to play with. The children enjoyed playing with both, although the robot frequently stalled and thus needed more adult intervention than did the crane. The children would frequently pull off Gizbar's wig, revealing the metal casing, wires and motors underneath. The robot would play peekaboo, and I was able to engage in shared game playing. This was not possible when engaging with children in the crane group. Objects that are more socially recognisable can be interacted with in socially interactive ways. The physical affordances of the robot supported our social interaction, those of the crane did not.

In light of my own reflections on the experiments with the robot and crane, I thought the children to be equally as interested in both types of objects. However, the children did appear more responsive when interacting with the robot, often imitating its movements, or touching its face and gazing in the direction of its eyes.

As I watched the replayed videos of our robot interactions, I was reminded of how many times the robot or its remote control ceased to operate. I often had to take my attention away from the child and return again to the computer to reset the controls. Re-watching the videos also reminded me of the cries and distress of the children. A child might be distressed by the sound, the light or some interruption to a routine they used to self-soothe. One such routine was demonstrated by Luke, a boy of about 5 years old: he would hum a tune, assort bricks in an order and then knock over a chair. Nothing could interrupt this routine. As my understanding of autism increased over time, I realised the rituals of these young children are ways of coping with a world they find distressing. Other children and parents I have met have also told me of how particular objects or ways of interacting can become central to self-soothing rituals.

At the end of the experimental process, the post-evaluation play took place. All the children were tested again on the same play protocols we had begun with. The children enjoyed the games and the toys. The children appeared more relaxed with me and engaged with me more, but this was regardless of whether they had been presented with the robot or the crane.

3.6 The Results

After the experiment was completed, the video-tape recordings were taken to an independent coder. I had no idea at this stage what the results of the experiment would be. I acknowledged that both objects interested the children, but the humanoid robot allowed social interaction to flow more recognisably between the child and me. Some weeks later I discovered the results—nothing had been found: no difference between how the toys impacted the children was observed. The research team put it down to the irregularity in each condition, that at times a child spent longer or shorter times with me, or the noise in the rooms outside could have interfered with the results. I accepted the study as a pilot—as it implies 'a run through before real thing'. But it occurred to me that this conclusion revealed something about the experimental process, specifically regarding the distortion of knowledge through scientific experimentation. In the experimental sciences, null results are not published. The experiment acts as a channel to reveal truth—but if the results reveal nothing of significance then this is omitted from the record. Pinch and Bijker (1984, pp. 399–441) describe this process as part of the 'social construction of facts' in which knowledge is produced according to implicated norms. Normative models of race, class, gender have all been shown by social studies of science and technology to shape disciplinary practices and direct dominant discourses (Helmreich 1998; Keller and Keller 2009; Collins 1998, pp. 494–516; Pinch and Bijker 1984, pp. 399–441).

If the research team had taken the null result seriously, if the negative results had not been explained away as a matter of 'interference' in the experimental process, then what would be left for the researchers? Would they be forced to recognise that their hopes to use robots as assistive devices for children with autism were ill-founded? Perhaps no difference was found because, regardless of the analogy often drawn, children with autism are not machines. There was no inherent affinity between the children and robots. This calls into question the effectiveness of robots in helping children with autism. Instead, this null result suggests that these children experience a complex sociality—an 'autistic sociality' that is set within human sociality, not somehow outside it or at the extreme end of some asocial continuum of an EMB.

Since the experimental study was completed and its results known, dozens of small-scale experiments developing robot therapy have been conducted. The results in this study showed no differential effect of the robot.

The robot did not miraculously make the children more social. It seems there was no special affinity between the children with autism and the mechanical robot after all.

Bibliography

Baron-Cohen, S., E. Ashwin, C. Ashwin, T. Tavassoli, and B. Chakrabarti. 2009. Talent in Autism: Hyper-Systemizing, Hyper-Attention to Detail and Sensory Hypersensitivity. *Philosophical Transactions of the Royal Society of London: Series B, Biological Sciences* 364 (1522): 1377–1383.

Bettelheim, B. 1959. Joey. *Scientific American* 200 (3): 116–130.

Bird, Geoffrey, Jane Leighton, Clare Press, and Cecilia Heyes. 2007. Intact Automatic Imitation of Human and Robot Actions in Autism Spectrum Disorders. *Proceedings: Biological Sciences* 274 (1628): 3027–3031.

Bourdieu, Pierre. 1980. Structure, Habitus, Practices. In *The Logic of Practice*. Cambridge: Polity.

———. 2011. The Forms of Capital (1986). In *Cultural Theory: An Anthology*, 81–93. Chichester: Wiley-Blackwell.

Breazeal, Cynthia. 2009. Role of Expressive Behaviour for Robots That Learn from People. *Philosophical Transactions: Biological Sciences* 364 (1535): 3527–3538.

Coeckelbergh, Mark, Cristina Pop, Ramona Simut, Andreea Peca, Sebastian Pintea, Daniel David, and Bram Vanderborght. 2016. A Survey of Expectations About the Role of Robots in Robot-Assisted Therapy for Children with ASD: Ethical Acceptability, Trust, Sociability, Appearance, and Attachment. *Science and Engineering Ethics* 22 (1): 47–65.

Collins, H.M. 1998. Socialness and the Undersocialized Conception of Society. *Science, Technology, & Human Values* 23 (4): 494–516.

Danforth, Scot, and Srikala Naraian. 2007. Use of the Machine Metaphor Within Autism Research. *Journal of Developmental and Physical Disabilities* 19 (3): 273–290.

Dautenhahn, Kerstin, and Iain Werry. 2004. Towards Interactive Robots in Autism Therapy: Background, Motivation and Challenges. *Pragmatics & Cognition* 12 (1): 1–35.

Feil-Seifer, D., and M.J. Matarić. 2005. *Defining Socially Assistive Robotics*. Proceedings of the IEEE 9th International Conference Rehabilitation Robot (ICORR 2005), June 28–July 1, Chicago, pp. 465–68. Piscataway: IEEE.

Foucault, Michel, and Jay Miskowiec. 1986. Of Other Spaces. *Diacritics* 16 (1): 22–27.

Frith, U. 1989. *Autism: Explaining the Enigma*. Oxford: Basil Blackwell.

———. 1991. *Autism and Asperger Syndrome*. Cambridge: Cambridge University Press.

Hacking, I. 2009. Autistic Autobiography. *Philosophical Transactions of the Royal Society of London: Series B, Biological Sciences* 364 (1522): 1467–1473.

Happé, Francesca, and Uta Frith. 2010. *Autism and Talent*. Vol. 364. Oxford: Oxford University Press.

Haraway, Donna Jeanne. 2003. *The Companion Species Manifesto: Dogs, People, and Significant Otherness*. Vol. 1. Chicago: Prickly Paradigm Press.

Helmreich, Stefan. 1998. *Silicon Second Nature: Culturing Artificial Life in a Digital World, Updated with a New Preface*. Berkeley: University of California Press.

Keller, Evelyn Fox, and Evelyn Fox Keller. 2009. *The Century of the Gene*. Cambridge, MA: Harvard University Press.

Kozima, H., C. Nakagawa, and Y. Yasuda. 2005. *Interactive Robots for Communication-Care: A Case-Study in Autism Therapy*. In IEEE International Workshop on Robot and Human Interactive Communication, ROMAN 2005, pp. 341–346. IEEE. August 2005.

Loh, Alvin, Teesta Soman, Jessica Brian, Susan E. Bryson, Wendy Roberts, Peter Szatmari, Isabel M. Smith, and Lonnie Zwaigenbaum. 2007. Stereotyped Motor Behaviors Associated with Autism in High-Risk Infants: A Pilot Videotape Analysis of a Sibling Sample. *Journal of Autism and Developmental Disorders* 37 (1): 25–36.

Ochs, Elinor, and Olga Solomon. 2010. Autistic Sociality. *Ethos* 38 (1): 69–92. https://doi.org/10.1111/j.1548-1352.2009.01082.x. http://dmu.summon.serialssolutions.com.proxy.library.dmu.ac.uk/2.0.0/link/0/eLvHCXM-w3V3NS8MwFA-KF1HE-cGmE3ZQL1Jpk7RJDh7GmOw2cB-gl9I2Ce6wI-awD_e99adKtnRM8Cz216aPNa3_5vffy3kOI4Eff28IETuHbAFtDEZGC-UajQCZCaU6khEMbz_bbC-tOo9cenW76AG7O_QfFd0GiKb78YDNvZ3ktcNsfD4Zr_jfM3uvZCCOAwrmNww-r_gATyiZVf0C9O1XV9VXLgL-JoKAIPBw6FlQNAClYlCWsISfiPL8HCne2y4hZO29Pud0guJbsioQGQj3KrZrUK9tbqtN4zWARJgd_5uEiEvzel0edyluVPauFNRrAKA1mjJSF2cSQmmKuaat-0vpdr5xPVTQ2TG7Ra7mQlBQMZn6ITZzp0ulblDbSnFmfoqNLp4us-cNUrtd9bav0CT5_64N_Bc2wsvA7rNPbMvNxHM51qHJE2oBpQMNBVgiXOqkzQNiRKG6MGEaLCmqUmmhvWTahXJLE3wJTpOTHrEIi_SKGUTdYSMhASZimhGE-6nWQhEVVLNfJxKzFqoXb50nKTGyZbly5hRzHAEnH73ZR7SoscR3H1n5yn-sCVSYhwvcezHjInQdDUyBeXzz7yFmlvjSMgj4FAMRNxWZ3g9YEvrLRT8ZVjP1bc3dR3yq7-JvkaHm9-pjQ40oIC-6QftyvvoGAFp00Q.

Peca, Andreea. 2016. Robot Enhanced Therapy for Children with Autism Disorders: Measuring Ethical Acceptability. *IEEE Technology and Society Magazine* 35 (2): 54–66.

Pinch, Trevor J., and Wiebe E. Bijker. 1984. The Social Construction of Facts and Artefacts: Or How the Sociology of Science and the Sociology of Technology Might Benefit Each Other. *Social Studies of Science* 14 (3): 399–441.

Pop, Cristina Anamaria, Ramona Simut, Sebastian Pintea, Jelle Saldien, Alina Rusu, Daniel David, Johan Vanderfaeillie, Dirk Lefeber, and Bram Vanderborght. 2013. Can the Social Robot Probo Help Children with Autism to Identify Situation-Based Emotions? A Series of Single Case Experiments. *International Journal of Humanoid Robotics* 10 (03): 1350025.

Pop, Cristina A., Sebastian Pintea, Bram Vanderborght, and Daniel O. David. 2014. Enhancing Play Skills, Engagement and Social Skills in a Play Task in ASD Children by Using Robot-Based Interventions. A Pilot Study. *Interaction Studies* 15 (2): 292–320.

Robins, Ben, Kerstin Dautenhahn, R. Te Boekhorst, and Aude Billard. 2005. Robotic Assistants in Therapy and Education of Children with Autism: Can a Small Humanoid Robot Help Encourage Social Interaction Skills? *Universal Access in the Information Society* 4 (2): 105–120.

Robins, Ben, Kerstin Dautenhahn, and Janek Dubowski. 2006. Does Appearance Matter in the Interaction of Children with Autism with a Humanoid Robot? *Interaction Studies* 7 (3): 509–542.

Scassellati, Brian. 1993. *Implementing Models of Autism with a Humanoid Robot.* Yale University. http://www.cs.yale.edu/homes/scaz/abstracts/1999/scaz3.pdf

Scassellati, Brian, Henny Admoni, and Maja Mataric. 2012. Robots for Use in Autism Research. *Annual Review of Biomedical Engineering* 14: 275–294.

Simut, R., J. Vanderfaeillie, B. Vanderborght, C. Pop, S. Pintea, A. Rusu, D. David, and J. Saldien. 2012. *Is the Social Robot Probo an Added Value for Social Story Intervention for Children with ASD?* In 7th ACM/IEEE International Conference on Human-Robot Interaction (HRI), 2012 HRI, pp. 235–236. IEEE. March 2012.

Tapus, Adriana, Andreea Peca, Amir Aly, Cristina Pop, Lavinia Jisa, Sebastian Pintea, Alina S. Rusu, and Daniel O. David. 2012. Children with Autism Social Engagement in Interaction with Nao, an Imitative Robot – A Series of Single Case Experiments. *Interaction Studies* 13 (3): 315–347.

Thill, Serge, Cristina A. Pop, Tony Belpaeme, Tom Ziemke, and Bram Vanderborght. 2012. Robot-Assisted Therapy for Autism Spectrum Disorders with (Partially) Autonomous Control: Challenges and Outlook. *Paladyn, Journal of Behavioral Robotics* 3 (4): 209–217.

CHAPTER 4

Echo and Narcissus and Pronoun-Reversal

4.1 The Myth of the Egocentric Child

We now turn to the themes of echolalia and pronoun reversal, two of the difficulties that social communication presents for children with autism. Social communication takes place through a series of social behaviours that include turn-taking, reciprocal interactions, pronoun use, joint attention and imitation. I want to explore what kind of sociality might be communicated by these behaviours. The term pronoun reversal is described by Leo Kanner (1943), writing

> Personal pronouns are repeated just as heard, with no change to suit the altered situation. The child, once told by his mother, "Now I will give you your milk," expresses the desire for milk in exactly the same words. Consequently, he comes to speak of himself always as "you," and of the person addressed as "I." Not only the words, but even the intonation is retained. (p. 244)

Or

> Her [Elaine] utterances are impersonal. She never uses the personal pronouns of the first and second persons correctly…She uses sentences just as she has heard them, without adapting them grammatically to the situation of the moment. When she says, "Want me to draw a spider," she means, "I want you to draw a spider." (p. 241)

K. Richardson, *Challenging Sociality*, Social and Cultural Studies of Robots and AI, https://doi.org/10.1007/978-3-319-74754-5_4

Kanner also referred to the behaviour of echolalia, the repeating back of words. Kanner distinguished between ***immediate*** ('Affirmation is indicated by literal repetition of a question') and ***delayed echolalia*** (p. 243). Echolalia would also be expressed by the child in the 'silent repetition of words' (ibid).

As speech terms, personal pronouns indicate the role of speaker and addressed; how might the reversing of these terms invite us to reflect on the autistic experience?

Personal pronouns are used by speakers to describe a role as addressee or speaker, and arithmetically signify number—as singular or plural identification with others. Personal pronouns provide an explicit means of crucial linguistic interpersonal framing of ontological experience without which we would be unable to include other perspectives in our experiences (Cromer 1974). Child developmental psychologist Jean Piaget [1896–1980] proposed that children underwent stages of development, identifying an 'egocentric' period in which the child can only see others from their own point of view (1964; Piaget 2013). Piaget contributed to seeing a child as primarily egocentric, unable to take into account an other, an idea challenged by Peter Hobson (1995), who found, depending on the language and the context, children as young as two-and-a-half could understand other perspectives and positions. The perspectives of children have also been explored by the work of Alice Miller, who offers an empathetic reading of childhood experience. Miller is particularly concerned to note how children are extremely responsive to their loved ones, while at the same time needing particular kinds of care as children because as children they are unable to meet their own needs and wants and need a non-exploitative relationship from their parents (1987).

The highly influential work of Jean Piaget shaped a generation of thinkers in child developmental psychology, and spawned a myth that children are primarily narcissistic and unable to consider the perspectives of others, including close kin. In Western philosophical tradition, the emphasis placed on the ego as the primary site of personhood informs developmental psychological notions about childhood, but also contributes to particular ways of understanding and making sense of autism. In 1913, a paper by Eugen Bleuler first used the term autism to describe schizophrenics as showing autistic behaviours and the inwardly, preoccupation with self and not others was reproduced in Kanner's (1943) paper. Kanner described the child Virginia:

> Not once have I seen her smile. *She retires within herself, segregating herself from others. She seems to be in a world of her own*, oblivious to all but the center of interest in the presiding situation...When others encroach upon her integrity, she tolerates them with indifference. There was no manifestation of friendliness or interest in persons. On the other hand, she finds pleasure in dealing with things about which she shows imagination and initiative. (my emphasis, p. 230)

Again the emphasis here for describing autism is on the withdrawn state. If typically developing children at this time were regarded as 'egocentric', it is no surprise that autistic children were regarded *as in a world of their own*. To what extent was the standard of egocentrism of the child—influencing the way that autism was viewed, understood, and seen by researchers—who operate within an individualist and egocentric tradition? To these issues we now turn.

4.2 Narcissus: From Myth to Sexual Pathology to Politics of the Individual

The term 'narcissism', according to Sigmund Freud, was derived from a clinical term used by Paul Nacke in 1899 to 'denote the attitude of a person who treats his own body in the same way in which the body of a sexual object is ordinarily treated – who looks at it, that is, to say, strokes it and fondles it till he obtains complete satisfaction through these activities' (1914, p. 73). The tale of Narcissus was popularised in psychoanalytic literature by Freud's *On Narcissism* (1914) to describe a pathological condition—primarily a sexual perversion. Narcissism, he wrote, 'arises through the drawing in of an object-cathexes as a secondary one, superimposed upon a primary narcissism that is obscured by a number of different influences' (ibid, p. 75). Freud's description further stated that narcissism was connected with a state of love: 'The highest phase of development of which object-libido is capable is seen as the state of being in love, when the subject seems to give up his own personality in favour of an object-cathexis' (ibid, p. 76). The emphasis here is on a condition in which the 'narcissist' pathologically gives up her/his identity to/for the other, or did not regard the other at all.

Primary narcissism is necessary for an infant, as infants are unable to meet their own needs, and need narcissistic gratification from their parents. The infant and child express magical thinking, and a sense of omnipotence evolves from this primary narcissism in which the child believes they can influence external reality. Secondary narcissism in children is withdrawal from others creating pathological states, 'they have taken as a model not their mother but their own selves. They are plainly seeking *themselves*[1] as a love-object, and are exhibiting a type of choice which must be termed "narcissistic"' (p. 88).

In the 1970s, American sociologist Christopher Lasch (1991) wrote *The Culture of Narcissism: American life in an Age of Diminishing Expectations* and argued narcissism was now a cultural force, moving beyond the individual pathology expressed by Freud. Lasch was concerned with the turn inward, advertising, consumerism and therapeutic salvation—'new therapies spawned by the human potential movement' (p. 9). Citing the work of Peter Marin, Lasch writes '"the individual will is all powerful and totally determines one's fate"; thus they intensify the "isolation of the self"' (ibid, p. 9). The individualist paradigm is associated with autonomy, agency and privacy, and has been central to the way in which Western political systems developed and subjects have understood the self since the Enlightenment. What Lasch objected to was not 'individualism' per se, but the narcissistic individualism of the new consumer middle class that went against collective and public-focused politics.

These readings by Freud and Lasch represent a move away from the tale of Narcissus in its original poetic description by the Roman poet Ovid in the *Metamorphoses* (1955). In the tale 'Echo and Narcissus' (Narcissus is not written about without Echo), a beautiful nymph, Echo, is cursed with the pain of only being able to repeat the words of the last things she hears. She finds a beautiful young man in the forest, Narcissus, and falls in love. But Narcissus, at his birth, had been predicted a disturbing fate: a seer told his mother, the nymph Liriope, that Narcissus would live to an old age only if he did 'not come to know himself' (Ovid and Innes 1955, p. 83). Narcissus could not love another. Thus, when he happened upon a lake and saw his own reflection, he fell in love with it. But a reflection is not a person, so Narcissus was deprived of human relationship, as was Echo. Their lives both ended in sadness:

> Many lads and many girls fell in love with him, but his soft young body housed a pride so unyielding that none of those boys or girls dared to touch

[1] Freud included 'perverts and homosexuals' in his category of 'people whose libidinal development had suffered some disturbance' (On Narcissism p. 88).

> him. One day, as he was driving timid deer into his nets, he was seen by that talkative nymph who cannot stay silent when another speaks, but yet has not learned to speak for herself. Her name is Echo, and she always answers back. (ibid, p. 83)

Narcissus was unable to see a human *other* as a separate being and engage in interpersonal interaction. In her sadness, Echo lost her body and only her voice remained. Narcissus was having a relationship only with his own self, which he had projected into the material forms around him. The form of the material was important, much like a doll, or effigy, or a humanlike robot. The water from the lake can bear a reflection and act as a mirror. Narcissus was relating only to himself through material goods that resembled him as a boy, but were not real. The watery reflection reduced Narcissus to an object with which no real intimacy or relationship could be developed:

> Narcissus, wearied with hunting in the heat of the day, lay down here: for he was attracted by the beauty of the place, and by the spring...he was enchanted by the beautiful reflection that he saw. He fell in love with an insubstantial hope, mistaking a mere shadow for a real body...Unwittingly, he desired himself, and was himself the object of his own approval, at once seeking and sought...But he could not lay upon himself. (Ovid and Innes 1955, p. 85)

Freud's punitive description of persons believed to have 'narcissistic' traits has given rise to an enormous literature on the subject. Narcissistic personality disorder, a clinical classification in DSM-IV, includes: 'A pervasive pattern of grandiosity (in fantasy or behavior), need for admiration, and lack of empathy, beginning by early adulthood and present in a variety of contexts, as indicated by five (or more) of the following':

1. Has a grandiose sense of self-importance (e.g., exaggerates achievements and talents, expects to be recognized as superior without commensurate achievements).
2. Is preoccupied with fantasies of unlimited success, power, brilliance, beauty, or ideal love.
3. Believes that he or she is "special" and unique and can only be understood by, or should associate with, other special or high status people (or institutions).
4. Requires excessive admiration.

5. Has a sense of entitlement, i.e., unreasonable expectations of especially favorable treatment or automatic compliance with his or her expectations.
6. Is interpersonally exploitative, i.e., takes advantage of others to achieve his or her own ends.
7. Lacks empathy: is unwilling to recognize or identify with the feelings and needs of others. (American Psychiatric Association 1994)

The DSM-5 thus uses pathologising language to describe different kinds of human subjectivities, collectively labelling them 'personality disorders'. There is no clear delineation between one disorder and another, and the 'symptoms' of one often overlap with those associated with another diagnosis. As a result, a person diagnosed with one mental health disorder can frequently be re-categorised as having another instead or in addition, as Brian McCabe explains in his biography in *Re-Thinking Autism*.

People with Asperger's have now been reclassified as autistic. The category of Asperger's entered DSM-IV, partly due to the clinical and campaigning work of American psychiatrist Lorna Wing (Timimi and McCabe 2016, p. 32). In 2013, DSM-5 abandoned the category of Asperger's Syndrome, combining its 'symptoms' with those of autism. Between 1994 and 2013, thousands of people have been diagnosed with Asperger's, and this has informed how they view themselves and experience the world. Some of these people were diagnosed in childhood and lived with the diagnosis throughout their lives. Others, such as the *Britain's Got Talent* singer Susan Boyle,[2] were diagnosed later in life (Boyle was diagnosed with the condition in 2013, in her 50s, the very year in which it was abandoned as a category in the DSM-5). What is to happen to people who have lived their lives, understanding themselves as having Asperger's? What about the thousands of Asperger's groups, websites and books on the subject? Are these no longer valid? Can one just cross out the word 'Asperger' and replace it with the word 'autism'? *The National Autistic Society* in the United Kingdom had this to say to concerning people with Asperger's:

[2] ITV Report. 28th April 2016. Susan Boyle's airport 'outburst' due to Asperger's Syndrome. [Online] Accessed 19.8.17 <http://www.itv.com/news/2016-04-28/susan-boyles-airport-outburst-down-to-aspergers-syndrome/>

> For many people, the term Asperger syndrome is part of their day-to-day vocabulary and identity, so it is understandable that there are concerns around the removal from DSM-5 of Asperger syndrome as a distinct category. Everyone who currently has a diagnosis on the autism spectrum, including those with Asperger syndrome, will retain their diagnosis. No one will 'lose' their diagnosis because of the changes in DSM-5.[3]

Timimi and McCabe (2016) note the overlap of 'symptoms' between the clinical diagnosis of autism and other personality disorders, including: hyper-sensitivities; lack of empathy; and failure to develop peer relationships appropriate to developmental level (p. 41). While Timimi and McCabe demonstrate the contradictory and problematic framing of human mental health in clinical categories, psychoanalysts worked from mythic conceptions of human beings. Psychoanalysts drew connections between autism and narcissism, pathologised by Nacke and Freud (Evans 2013; Grinker 2008; Bettelheim 1967, 1959). Frances Trustin for example, saw the pathology as a response to traumatic circumstances and tied 'infantile autism' to 'infantile narcissism' (Symington 1993). Trustin believed the autistic child uses the object (here, referring to a person) in pathological ways rather than through typical interpersonal reciprocal interactions that rely on mutuality and interdependence. Instead the autistic child used the person 'for a manipulative, parasitic relationship in which the object is to be seduced and controlled so as to allow the autistic/narcissistic subject to remain omnipotent and protectively encapsulated. Like the autistic child, the narcissist hates object relations[4] but is stuck with them and so has to manipulate them to their pathological needs – and has to suffer the consequences' (Grotstein 1993, p. xvi).

Autism has also been connected with narcissism in biomedical psychiatry and clinical psychology. This can especially be seen in the idea that people with autism are systematisers and hence lack empathy. This is associated with the fictional character Narcissus who could not recognise the experience of others. Like Narcissus, autistic persons are viewed as being

[3] The National Autistic Society: Autism profiles and diagnostic criteria [Online] Accessed 19.8.207 < http://www.autism.org.uk/about/diagnosis/criteria-changes.aspx>

[4] Object relations is a psychoanalytical concept developed by significant figures including Melanie Klein and Donald Winnicott to describe how 'objects' (people and relationships with them, particularly the mother) are introjected (taken inside the growing infant as categories). Starting in infancy, the infant intrapsychically internalizes relations with the mother, or other significant caregiver relationship (Sayers 1992).

outside of interpersonal relationships and locked inside a 'systemising' asocial universe. The pathologising language of these fields is clear in the titles of their papers. For instance: '*Can emotion recognition be taught to children with autism spectrum conditions?*' (Baron-Cohen et al. 2009, pp. 3567–3574), '*The Empathy Quotient: An Investigation of Adults with Asperger Syndrome or High Functioning Autism, and Normal Sex Differences*' (Baron-Cohen and Wheelwright 2004) and, most extreme, '*Zero degrees of empathy: A new theory of human cruelty*' (Baron-Cohen 2011). This language is damaging to autistic people according to Runswick-Cole (2016), Cohen-Rottenberg (2012) and Robinson (2014).

4.3 Echolalia, Pronoun-Reversal and I-You Interrelatedness

The themes from Ovid's *Echo and Narcissus* are threaded through particular psychoanalytical and biomedical models of autism. The name of Echo, the mythic nymph from *Metamorphoses,* forms the root of 'echolalia'—the echoing back of words to the speaker—a condition associated with autism. Why are Echo and Narcissus, such tragic figures, used as reference points for helping make sense of autism? Can there be another reading of this tale in the context of autism? And how does this relate to the use of robots in an attempt to elicit socialness from autistic children? How is this socialness conceived of and measured? Is it a smile or a frown, turn-taking, imitation or responsiveness?

Muskett (2016) is concerned by how particular kinds of phenomena, including speech patterns like pronoun-reversal or echolalia, are treated as 'signs' of autism. He believes that decontextualised sets of rules for language and communication do not take into account several important points for consideration including: power relationships; the status of the speaker and listener; the, age, language and expressive capacities of the speaker and listener; intention; cultural norms; emotions; wider context.

Despite these variables, guides to appropriate language-communication are presented like normative formulae for interpersonal communication. In reality, no such formula exists as each encounter is spontaneous and grounded in the specifics of each context. Muskett writes, 'decontextualized description represents a common feature of psychiatric classification, and in disciplines outside of psychiatry there is a comparable long history of individualised and asocial theorising of language and communication'

(p. 305). Language, in Western psychology, 'has traditionally been theorised as a window into a private realm of mental states' (ibid). The underlying individualist paradigm, which seeks to use the individual as the start and end point, blurs the dynamic roles of language and communication. It achieves this by presenting them as a static formula, rather than an iterative, spontaneous, dialogical processes, that at times is chaotic, indeterminate, and confusing.

Language-communication serves an important role in helping the speaker to communicate their needs, wants and desires to the listener, and for the listener to acknowledge that those needs, wants and desires have been heard. But such precise and synchronised understanding between speakers and listeners occurs primarily in the context of the ongoing encounters which help to build intimacy and trust over a period of time. Again, as Martin Buber wrote, presence in dialogue and encounter is fleeting, coming and going, returning to the I-It form of unknowing. Discontinuity, breaks and interruptions form the basis of all human speech and language encounters, not just that of children with autism or speech and language difficulties.

Personal pronouns—I, you, he, she, it, we, they, me, him, her, us, and them—indicate positionality; the subject pronoun I is tied to the speaker, whereas you is the addressed. The I-you is reversible. The addressed, in the role of you, swaps roles with the speaker, I, in dialogue.[5] However, there are also other forms of speech including we-her/him/it or us-they/me-you.

Martin Buber, in his *I and Thou* (1937), argued that the speaker is also tied to the addressee, writing:

> The attitude of man is twofold, in accordance with the twofold nature of the primary words which he speaks. The primary words are not isolated words, but combined words.
>
> The one primary word is the combination I-Thou.
>
> The other primary word is the combination I-It; wherein, without a change in the primary word, one of the words He and She can replace It. (p. 2)

[5] Object here should not be read as a material inorganic artefact, but as another person, separate and existing independently of the I, which looks back at the I, as a You. The term 'object' to describe another person is not ideal, in fact it is used widely in psychoanalytical studies 'object-cathexis' (to describe using another like one self) or 'object-relations' which regards the way in which people related to each other transpersonally.

Buber continues 'There is no I taken in itself, but only the I of the primary word I-Thou and the I of the primary word I-it' (p. 3). If there is always a you implied in the speaking of I—which kind of other is invoked? Buber's writings are mystical, and primarily concerned with the relationship with God, He in the Third Person who he calls the Eternal Thou, but through this narrative he invites us to consider human relationships and what kind of address is spoken to others. Much like the narcissistic framework, the I-it form is also preoccupied with the I-experience over and above the I-thou experience 'I perceive something. I am sensible of something. I imagine something. I will something. I feel something. I think something. The life of human beings does not consist of all this and the like alone. This and the like together establish the realm of It' (p. 3). Buber's poetic expression of interpersonal relationship is described here: 'If I face a human being as my Thou, and say the primary word I-Thou to him, he is not a thing among things, and does not consist of things' (p. 7). The power of Buber's phenomenology is the invitation is to be with other humans, not as objects, as a means of gratifying the I, but of relating to them as in their complete existence. It means taking account of what the other might be experiencing, thinking and feeling as unique.

According to Smith in his discussion of Buber 'The world of objects or things, on the other hand, presupposes a single centre of consciousness [the egocentric individual of modernity], one subject, an I which experiences, arranges, and appropriates[6]' (Smith cited in Buber 1937, p. vi). The absence of a you or rather transformation of a you into an It in modern individualism has profound consequences and dangers for the way our knowledge is produced. If the you becomes absent or denied, what are the results of this? I-thou is about meeting the you as they are in the moment, entering into their subjectivity without losing one's own. By contrast, the I-it combination is about relating to the addressee as an It, a thing, an object. 'It' has existence only in the I's projection. I-it relationships have underlain—and continue to underlie—slavery, colonialism, torture, racism, sexism and cruelty towards children.

The I-thou form can be presented in a more familiar, but equally powerful form as the I-you. I-you are interrelated speech forms and co-exist and are co-experienced by human beings depending on which speech or listener role they take in conversation. The I-you is learned, according to Stawarska (2009) in face-to-face dialogues between infant

[6] Ronald Gregor Smith, Introduction I and Thou (1937).

and parents or other significant caregivers, and therefore cannot exist outside of human interpersonal relationship. The I-you represents speech utterances, but is also interpersonal, and cannot come from objects into the lives of living beings, as the I-you grows inside intimate interpersonal ontological experience between infants and their caregiving family. Peter Hobson (1995), Professor of Developmental Psychopathology at University College London and an expert on autism, took the idea of I-thou and I-it from Buber and reworked them in his cognitive model of autism, suggesting that autistic children have deficits in I-Thou relatedness expressed by mutual gaze, co-ordinated and shared attention, turn-taking, or appropriately imitate relationship and proposed instead 'thing-relatedness' (p. 196). Hobson even proposed that for an autistic child I-it was a 'relatively successful…developmental pathway' (p. 197).

The origin of the separation of the I from the you occurred thanks to developments in Western philosophy and psychology that privilege egocentrism. In the egocentric tradition the 'I' was denoted as separate from the 'you' to describe a 'discourse-independent reference (such as the mind of thinker)' or cast it grammatically as a 'substantive noun, typically prefaced with a definite article' (p. 18). Beata Stawarska, author of *Between You and I: Dialogical Phenomenology* writes: 'the modern turn inward is facilitated…by a focus on the first-person perspective as the purported point of entry into the rational mind and by a focus on the personal stance of the "I," over against the personal stance cast in the other personal pronouns (notably "you" and "we")' (Stawarska 2009). The egocentric tradition gave precedence to first-person subjectivity, over and above others. The upshot is an egocentric tradition that is missing a 'you'; the 'you' is considered only as an object of the 'I'. This confusion has distorted how humans brought up in this tradition interrelate. Put another way, the philosophy of Husserl, Descartes, Locke and Kant all lack a 'you'. The 'you' is philosophically, but not empirically, absent: the 'first-person stance is entangled with the second-person address' (ibid, p. 19).

4.4 Robots to Humans as the Lake Is to Narcissus

Having emphasised the myth of Echo and Narcissus and its interpretation by psychoanalysis, political scientists and autism researchers, I now turn my attention to experiments that explore robots' potential to encourage children with autism develop social skills. The myth of Echo and Narcissus are used to make sense of autism, specifically the view that autistic persons are uninterested in others, reject relationship with others, and prefer the company of 'things', interpreted by robotics scientists to mean interest in machines and robots over engagements with other humans. To accept this proposition that autistic children or adults are less interested in people than things, the egocentric model is privileged above all others. As we will see in the next chapter on attachment and I-you interrelatedness, there is no One in the I, humans are a social species whose lives depend on interpersonal connectedness and relationality. Therefore people with autism are *with others*, in co-experience with others, as are all humans and living species. Moreover, the view that people with autism lack empathy and do not want friendship has been challenged from a number of studies and personal accounts by autistic people themselves.

Rosqvist et al. (2015) explore the meaning of friendship for autistic adults and draw on an article written by autistic people for a Swedish advocacy magazine *Empowerment*. The authors reject the naturalised 'construction of autism as a deficit characterised by disordered emotional state and lack of sociality' (p. 1). Using the lens of neurodiversity, the authors propose that autism is a neurological difference rather than a deficit. The authors still describe neurodiversity as being *inside* biomedical and neurological models, even though a genetic or neurological basis to autism has yet to be discovered (Runswick-Cole et al. 2016). That said, neurodiversity networks of autistic adults facilitate friendship and sociality, somewhat countering the 'naturalised' assumptions that autistic people are not interested in friendship or other people. Robinson (2014) developed participatory research methods involving adults with Asperger's syndrome. This led to a practice of mutual understanding and a Triad of Understanding:

Me Understanding Myself
Me Understanding Others
Others Understanding Me. (p. 226)

Robinson describes these as interrelated and iterative. Robinson emphasises the importance of 'autistic spaces', where autistic people can come together and explore their autism—a theme explored by Rosqvist et al. (2015). For friendship, self-awareness and awareness of others to develop fully, autistic adults must have spaces in which they can be as they are without 'professionals' trying to fix them or pathologise their behaviours. Robinson's work involves coming alongside and developing alongside autistic adults. Autism acceptance is important.

Groups that are designed to support autistic young people and adults and provide a non-threatening and accepting environment are extremely beneficial to help independence and self-acceptance. I have also found this to be the case during my own time spent with adults with Asperger's and the most popular group activity was 'social skills'. The young adults enjoyed learning about social interaction. Several group members told me about their difficulties making friends at school and being bullied. John and Cat had told me they did 'not fit in at school' among their peers and felt lonely and distressed. The group gave them an opportunity to meet like-minded people and have fun. Sometimes adults might also say they are not interested in friendship or close relationship as a self-defence strategy. Another example of these supportive environments is the pioneering *People who Hear Voices*[7] networks that bring people together to discuss the hearing of voices, and explore what these voices mean.

Psychodramatist, John Casson (2004) works with people who hear voices by using drama as a way of engaging people with psychosis. He has worked with adults with severe psychosis, primarily schizophrenia, using psychodramatic techniques to explore their voices. Adults that gained more awareness over their voices were able to engage with the voices, to dialogue with them. Adults who hear voices and present these symptoms to a medical doctor will be referred to a psychiatrist and applied a label. The psychiatrist's primary goal, using medications such as Lithium, a powerful drug, is to stop the person from hearing voices. Psychodrama and other humanistic approaches to human suffering try to engage with what the human being is actually experiencing and support the person in developing self-awareness over their inner dialogues.

Drama as a therapeutic form provides a means for people to explore their alterity or difference and push the limits of their everyday realities. The power of theatre was also known to the ancients; the first theatres were built next to temples because of the respect for the transcendental nature of theatrical experience. It is no surprise that the ancient Greek theatre of Dionysus Eleuthereus was used for festivals in honour of the god Dionysus. In some theatre styles, the actor leaves herself to be temporarily Other in a performance for an audience. The audience must also participate in this Otherness by buying into the performance, suspending

[7] Hearing Voices Network, for people who hear voices, see visions or have other unusual perceptions. Accessed 23.8.17 < https://www.hearing-voices.org/>

reality. The performance is liminal, but the action and experience of that action by actors and audience is located in time, space and place.

I proceed to draw on the ways in which echolalia was part of my interactions with some of the children during the time of the experiment in order to think about what a child might experience when they are given a robot and asked to interact with it. The robot in this experiment could not speak, and so could not echo. The only sound produced by the robot was a faint buzz of the computer or the whirring sounds of motors moving when it was triggered to move via a remote control, held in my hands or the hands of the child. This robot was developed in part with the intention of encouraging learning through imitation: if the robot moved, the child would be encouraged to play a game to imitate the robot, then I would let the child control the robot, and if the robot moved, I would imitate it.

Learning theories (Bowlby 2008) emphasise the role of imitation as one of the pillars of learning strategies. Imitation operates alongside the development of other learning skills, and thus, interpersonal social interactions appear pivotal in how people learn.

In autism, social-communicative interaction and imitation skills are difficult. Thus the question is not whether a child with autism can imitate the actions of a robot in a particular game-playing setting, but do the imitation games learned through playing with the robot have long-term benefits when no longer in the robot's presence? Can the child take what is learned in the therapeutic settings and generalise the experience to other contexts?

Contrary to reports about the effectiveness of robots, in our pilot experiment exploring the effectiveness of a humanoid robot for children with autism spectrum conditions we found no difference in the data between the control group and the experimental group: children who interacted with the robot showed no empirically observable difference in behaviour from those who interacted with the crane. In this case-study there was no effect from interacting with a robot. These results were never published because empirical researchers need not publish null results. Moreover, given that the children only had objects (toys, a crane or robot) as the focal point of their interactions, was the experiment not starting from an assumption that is problematic? Moreover, relationships between people develop along different lines than interaction with objects. People form relationships based on intimacy, trust and experience and cannot be correlated with inanimate objects. Robots for children with autism derive from an idea that children with autism prefer things over other people. It is this basic assumption, which implicitly contains the egocentric perspective, that is problematic. The egocentric and individualist perspective continually guides what adults believe children to be, autistic or not.

The unique pronoun use of autistic children was described in the paper, *I, You, Me and Autism: An Experimental Study* (Lee et al. 1994, pp. 155–176), which notes that it is rare for other children to use pronoun reversal although it is not exclusive to autistic children—blind children also display pronoun reversal—but their findings reveal that children with autism did not consistently avoid personal pronouns and in some cases were more likely that a comparative group of children to use the pronoun I. The findings also reveal difficulty in showing links between echolalia and pronoun reversal, and the correct usage of pronoun might assist a child with echolalia to speak. 'A possible partial exception here is that autistic subject's tendency to respond to a question such as "Who can see the X?" with "I can see the X" might be counted a form of mitigated echolalia' (pp. 172–173).

To conclude I will explore an interaction between myself and Bea at the school. Bea was around 8 years old and her parents had agreed for her to be part of the experiment. As all the children were randomly allocated into the crane or robot groups, Bea was assigned a place in the crane group. Bea used to enjoy coming into the class and playing with the crane, or at least I thought so. She showed no protest at coming to the sessions, nor any protest at the play. I usually had to let her know that time was up, and she could go back to her classroom.

Bea's speech to me had echolalia elements. In the experiment, I tried to be as socially interactive with the children with the crane as with the children with the robot. It was easier to be socially interactive in the presence of a child with the robot because I could use pronouns and name differently, such as 'look at the robot Gizbot'. I frequently referred to the robot as 'him' because he looked like a small boy of around 3 or 4 years old (though I was told Gizbot had no gender, he was dressed in clothes that were fitting to a young boy.)

I used to engage Bea by saying 'do you like to play with the crane?' or 'what can the crane do' or 'shall we put some blocks into the crane's basket?' Bea often carried out the correct actions (she put blocks in the basket) but simultaneously repeated back to me the words that I spoke to her. Most of all she seemed to enjoy playing with the remote controls of the crane. I too imitated her speech in return. Like the poet proclaimed: 'Echo still had a body then, she was not just a voice: but although she was always chattering, her power of speech was no different from what it is now. All she could do was to repeat the last words of the many phrases that she heard' (Ovid and Innes 1955, p. 83).

This back and forth of imitation between us was considered. Bea picked up my words and echoed them back to me; I slightly modified her words as I returned them to her, and she echoed them back to me. This sort of use of language is described in the literature as 'non-functional' since it does not aid or promote communication. But if it is all a child can do, then it arguably does have a function related to interpersonal interaction. It need only be seen as non-functional when understood through an egocentric model of communication. Communication, seen in the egocentric individualist tradition is nothing more than mechanical exchanges of signals that can be written down as guidelines or formulas. The aliveness of the speech-acts of an autistic child, why they echo, or pronoun-reverse are lost on the addressee.

Bibliography

American Psychiatric Association. 2013. *Diagnostic and Statistical Manual of Mental Disorders (DSM-5®)*. American Psychiatric Pub.

Baron-Cohen, Simon. 2011. *Zero Degrees of Empathy: A New Theory of Human Cruelty*. Vol. 30. London: Penguin.

Baron-Cohen, Simon, and Sally Wheelwright. 2004. The Empathy Quotient: An Investigation of Adults with Asperger Syndrome or High Functioning Autism, and Normal Sex Differences. *Journal of Autism and Developmental Disorders* 34 (2): 163–175.

Baron-Cohen, Simon, Ofer Golan, and Emma Ashwin. 2009. Can Emotion Recognition Be Taught to Children with Autism Spectrum Conditions? *Philosophical Transactions: Biological Sciences* 364 (1535): 3567–3574.

Bettelheim, Bruno. 1959. Feral Children and Autistic Children. *American Journal of Sociology* 64: 455–467.

———. (1967). *Empty Fortress*. New York/London: Free Press; Collier-Macmillan.

Bowlby, John. 2008. *Attachment*. Basic Books.

Buber, M. 1937. I and Thou (trans: Smith, R.G.). Edinburgh, Clark.

Casson, John. 2004. *Drama, Psychotherapy and Psychosis: Dramatherapy and Psychodrama with People Who Hear Voices*. New York: Routledge.

Cohen-Rottenberg, Rachel. 2012. *Deconstructing Autism as an Empathy Disorder: A Literature Review*. Autism and Empathy.

Cromer, Richard. 1974. The Development of Language and Cognition: The Cognition Hypothesis. In *New Perspectives in Child Development*, ed. B. Foss, 184–252. London: Penguin.

Evans, Bonnie. 2013. How Autism Became Autism: The Radical Transformation of a Central Concept of Child Development in Britain. *History of the Human Sciences* 26 (3): 3–31.

Grinker, Roy Richard. 2008. *Unstrange Minds: Remapping the World of Autism*. New York: Da Capo Press.

Grotstein, J.S. 1993. Melanie Klein in Perspective. *Psyccritiques* 38 (4): 420–422.

Hobson, R. Peter. 1995. *Autism and the Development of Mind*. Psychology Press.

Kanner, L. 1943. Autistic Disturbances of Affective Contact. *Nervous Child* 2 (3): 217–250.

Lasch, Christopher. 1991. *The Culture of Narcissism: American Life in an Age of Diminishing Expectations*. New York/London: W.W. Norton & Company.

Lee, Anthony, R. Peter Hobson, and Shulamuth Chiat. 1994. I, You, Me, and Autism: An Experimental Study. *Journal of Autism and Developmental Disorders* 24 (2): 155–176.

Miller, Alice. 1987. *The Drama of Being a Child: And the Search for the True Self*. London: Virago.

Muskett, Tom. 2016. Examining Language and Communication in Autism Spectrum Disorder – In Context. In *Rethinking Autism: Diagnosis, Identity and Equality*, ed. K. Runswick-Cole, R. Mallett, and S. Timimi, 300–316. London/Philadelphia: Jessica Kingsley Publishers.

Ovid and Mary M. Innes. 1955. *The Metamorphoses of Ovid*. Harmondsworth: Penguin.

Piaget, Jean. 1964. Part I: Cognitive Development in Children: Piaget Development and Learning. *Journal of Research in Science Teaching* 2 (3): 176–186.

———. 2013. *The Construction of Reality in the Child*. Vol. 82. London/New York: Routledge.

Robinson, Jacqueline. 2014. *Participatory Research with Adults with Asperger's Syndrome: Using Spatial Analysis to Explore How They Make Sense of Their Experience*. Unpublished PhD thesis. Leicester: De Montfort University.

Bertilsdotter Rosqvist, H., Brownlow, C. and O'Dell, L., 2015. 'An Association for All'—Notions of the Meaning of Autistic Self-Advocacy Politics within a Parent-Dominated Autistic Movement. Journal of Community & Applied Social Psychology, 25(3), pp. 219–231.

Bertilsdotter Rosqvist, H., Brownlow, C. and O'Dell, L., 2015. " What's the point in having friends?" reformulating notions of friends and friendship for adults with autism. Disability Studies Quarterly, 35(4) [Online].

Runswick-Cole, K. 2016. Understanding This Thing Called Autism. In *Re-thinking Autism: Diagnosis, Identity and Equality*, ed. K. Runswick-Cole, R. Mallett, and S. Timimi, 19–29. London/Philadelphia: Jessica Kingsley Publishers.

Runswick-Cole, K., R. Mallett, and S. Timimi. 2016. Understanding Introduction. In *Re-thinking Autism: Diagnosis, Identity and Equality*, ed. K. Runswick-Cole, R. Mallett, and S. Timimi, 7–18. London/Philadelphia: Jessica Kingsley Publishers.

Sayers, Janet. 1992. *Mothering Psychoanalysis: Helene Deutsch, Karen Horney, Anna Freud and Melanie Klein*. Harmondsworth: Penguin.

Smith, Ronald Gregor. 1937. Introduction. In *I and Thou*, ed. M. Buber. Trans: R.G. Smith. Edinburgh: Clark.

Stawarska, Beata. 2009. *Between You and I: Dialogical Phenomenology*. Athens: Ohio University Press.

Symington, Neville. 1993. *Narcissism: A New Theory*. London: Karnac Books.

Timimi, S., and B. McCabe. 2016. Autism Screening and Diagnostic Tools. In *Re-thinking Autism: Diagnosis, Identity and Equality*, ed. K. Runswick-Cole, R. Mallett, and S. Timimi, 30–48. London/Philadelphia: Jessica Kingsley Publishers.

CHAPTER 5

Attachment Theory and Autism

5.1 Bonds and Ties

Autism is not a concept fixed across time and space, and nor is the concept of parenting or childhood. The way in which childhood is considered and experienced has varied over time. Some writers, such as Edward Shorter, have highlighted the social construction of motherhood, mothering and infant-child relationships. Moreover, studies of European history indicate the family unit as we know it, along with attitudes towards children, have shifted considerably. Contemporary European and American attitudes towards childrearing, codified in law, restrict and prohibit behaviours towards children that were once commonplace and permissible. In *The Making of the Modern Family* (1975) Edward Shorter was concerned to show how intimacy, love and spontaneity grew out of modernity, writing:

> The little band of scholars that for some time now has been arguing that in traditional society mothers didn't love their children very much has met with stark incredulity. (1975, p. 169)

Did mothers love their children before modern society invented 'mother-love'? According to the social constructivists led by Philippe Aries, they did not. Shorter writes that this group of thinkers, 'with the aid of portraits and family reference books…concluded that small children were seen in the Middle Ages as creatures apart from people' (Shorter 1975, p. 169).

K. Richardson, *Challenging Sociality*, Social and Cultural Studies of Robots and AI, https://doi.org/10.1007/978-3-319-74754-5_5

Indifference and violence were regular penalties for a child born during that epoch. This included rocking children violently, laying children in soiled swaddling for hours on end, leaving children alone (or with younger siblings) during the day, or taking them to work where they experienced dangerous working conditions. According to Shorter, 'wakeful children were commonly *knocked* into the sleep of sensibility' (p. 170, original emphasis) by a contraption attached to the cot above the child's head!

Despite experiencing a tyranny of violence, Shorter argues 'it is not an inverse indication of affection. Children were brutalized by the daily routines of life as much as by savage outbursts of parents' (p. 170). What led parents to abandon children during this period was 'desperate poverty' (p. 174). The practice of wet-nursing (sending children to be weaned on the commercial market) was a common practice in France in urban centres. Poorer women would send away their own children and take in infant 'nurslings' from wealthier families, suggesting a lack of maternal attachment to their own children (pp. 178–181). Shorter argues that the process of maternal attachment can be seen in the mid-1700s, firstly among the middle classes (p. 182), when mothers began to nurse their own babies. A census taken in 1796 showed that '98 percent of children on the breast were being fed by their mothers' (p. 184).

Lawrence Stone, author of *The Family, Sex and Marriage in England 1500–1800* contrasts the *open lineage family* (from the sixteenth century and a millennium before) with *affective individualism*. In the open lineage family system:

> the early sixteenth-century home was neither a castle nor a womb. Lacking firm boundaries, it was open to support, advice, investigation and interference from outside, from neighbours and from kin; an internal privacy was non-existent. Inside the home the members of the nuclear family were subordinated to the will of its head, and were not closely bounded to each other by warm affective ties. (Stone 1979, pp. 6–7)

Stone charts the development of *affective individualism,* central to intimate affective relationships, organised through:

> Intensified affective bonding of the nuclear core at the expense of neighbours and kin; a strong sense of individual autonomy and the right to personal freedom in the pursuit of happiness; a weakening of the association of sexual pleasure with sin and guilt, and a growing desire for physical privacy. (ibid, p. 8)

Affective relations between parents and children slowly evolve in Stone's account of child-rearing attitudes and practices in England and France between 1666 and 1800. According to Stone, modern parents were encouraged to feel less affectionate towards their children. For instance, John Locke, who noted a growing 'affectionate approach to the problems of child-rearing' also warned parents against 'fondness' of a child (p. 407). Records in the sixteenth-century indicate that among the upper-classes, fathers felt little compassion for their children, with philosopher Michel de Montaigne [1533–1592] writing 'I have lost two or three children in infancy, not without regret, but without great sorrow' (cited in Stone, p. 105). There is little evidence from the period of mourning the death of infants, and the longer a child 'lived the more likely an affective bond would develop between it and parents' (ibid, p. 106). While parent-child affective relations were often lacking during this period, siblings sometimes demonstrated great affection. Stone remarks: 'there is reason to suspect that in the sixteenth and seventeenth centuries the brother-sister relationship was often the closest in the family. In the eighteenth century, when family bonding generally became much closer, there are still signs that brother-sister ties were particularly intimate' (ibid, p. 115). Children in the early modern family were not seen as 'fully human. They were smelly, noisy little creatures driven by instinct, not reason, incapable of communicating except at the animal level, and with only the most precarious hold on life' (Stone 1979, p. 166). During these times, nobody seemed to care much for infants and children, except the church (baptism of children was a priority). During these times, children were *othered*, seen as distinct and inferior. While there were some changes in affective ties between parents and children from the late 1600s, it was not until the eighteenth century that intimacy and involvement in children's lives appears to have become more important.

In some cases the roles of the 'mother' could be passed onto a new bride or older female if she was to die. The same is true of the role of the father, which if he died could be occupied by male members of the family (Marvick 1974, p. 288).

It was not until the ninetieth century that more systematic concern for the welfare of the child was developed across the class spectrum. This concern was expressed in harsh discipline of the child. By the twentieth century, the theme of 'love' and affective bonds had become conventional in how parent-child relations were understood. These were particularly celebrated in the form of 'mother-love' of the child.

The social construction of 'mother-love' has operated in multiple ways, acting as a legitimating strategy for establishing the division of labour: males would work while females would take care of their male spouse and children. It is important to distinguish between the role of parents, which encompasses the duality of caregiving females and males in a child's life, and mothering, which is a focus on the specific importance of the mother in the life of the child. As the mother was also a source of food for an infant, this relationship was of primary importance for a child's survival, but as shown, nursing of infants could be contracted out to other lactating women for breast-feeding. It was not necessarily important that the biological mother's milk was ingested by the infant, but in property owning societies it was important that the infant was the father's offspring for the maintenance of patriarchal relations (Engels 2010). While the shift to the mother as an important figure in the child's life increased her importance, this was accompanied by a strengthening of patriarchal power over women and children (ibid).

5.2 Mother-Love to Mother-Hate

A theory of different attachment patterns was developed in the 1940s and 1950s by psychologist John Bowlby. Bowlby explored how parents'—particularly the mother's—relationship with their child contributed to the child's ontological development and welfare. Bowlby was keen to introduce the concept of 'love' in looking at how affective relationships were of central importance to an infant and child's development. Bowlby (1982[1969]) was commissioned by the World Health Organization (WHO) to carry out a study on children coping with crises. He wrote:

> What is believed to be essential for the mental health is that the infant and young child should experience a warm, intimate and continuous relationship with his mother (or permanent mother-substitute) in which both find satisfaction and enjoyment. (1982, p. 12)

He also added: 'From empirical observation we suggested that "the young child's hunger for his mother's love and presence is as great as his hunger for food", and that in consequence her absence inevitably generates "a powerful sense of loss and anger"' (Bowlby 1982, p. 2). This is in contrast to the account of the early modern parent's frequent absence or indifference to children. Maternal love, as it was known, was not always present with the birth of a child and a mother could reject the child with social approbation (Marvick 1974, p. 289). Bowlby considered his approach to attachment 'scientific' and combined psychoanalytic perspectives with developmental psychology, ethnology and evolutionary psychology. Attachment theories rest on the idea of 'instinct' which underlies Freudian psychoanalysis, as Bowlby believed he had found this evidence from attachment studies:

> During the half-century since Freud sought a well-based theory of instinct and lamented his inability to find one, striking progress has been made… three empirically based sciences: ethnology (the study of animal behaviour by zoologists), experimental psychology and neurophysiology – Freud's own first love. (1982, p. 61)

Attachment explores species-specific and cross-species attachment styles, behaviours and patterns, and the consequences of disruption to these attachments by abuse, warfare, violence, poverty, and neglect. A child's positive experience of an intimate encounter, according to developmental psychologists, supports the child in later life to develop friendships and pair-bonding partnerships (Bowlby 2008). Writing as a challenge to Melanie Klein's theory of 'object-relations', Bowlby was keen to establish a new vocabulary of relatedness, using terms such as 'bond' or 'attachment figure' or 'attachment' instead of analytic terminology (ibid, p. 177). Object-relations was a term informed by a theory of instinct in which 'the object of the instinct is defined as "the thing in regard to which or through which the instinct is able to achieve its aim"' (Freud cited in Bowlby 2008, p. 177).

The role of the 'mother' became increasingly important alongside the rise of affective individualism (Stone) and the cultivation of the division of labour and the private sphere of the nuclear family.

As Janet Sayer, author of *Mothering Psychoanalysis* (1992) writes, the shift in emphasis to the mother from the father in psychoanalytic theory marked the move away from the patriarchal and phallocentric versions of psychoanalysis that had dominated since Freud. She points to the shift as represented by the arrival and remodelling of theories of early childhood in the work of Helene Deutsch, Karen Horney, Anna Freud and Melanie Klein (p. 3). The 'mother' and her role in early childhood had not been as central to the first psychoanalytical theories.

Freud wanted to put psychoanalysis on a par with medicine (Holmes 1993), a goal Bowlby believed he achieved with attachment theory. Bowlby combined the analytical insights of the psychoanalytical school—Winnacott, Klein, Anna Freud—with the studies of ethnology and evolutionary psychology. Mary Ainsworth (Ainsworth 1978) continued the work of Bowlby in her Strange Situation Experiment. In the experiment, a one-year-old child is brought into a room with an experimenter for 20 minutes. Then the mother leaves the room for three minutes and returns. Then the mother and experimenter leave the room and the child alone. Then after three minutes the mother returns. Ainsworth measured and observed the behaviours of children during these events. Ainsworth developed a model of attachment based on how much distress or indifference or comfort the child displayed when the mother left, and returned. The schema is as follows:

1. Secure attachment ('B')These infants are usually (but not invariably) distressed by separation. On re-union they greet their parent, receive comfort if required, and then return to excited or contented play.
2. Insecure-avoidant ('A') These children sow few overt signs of distress on separation, and ignore their mother on re-union, especially on the second occasion when presumably the stress is greater. They remain watchful of her and inhibited in their play.
3. Insecure-ambivalent (insecure-resistant) ('C') They are highly distressed by separation and cannot easily be pacified on re-union. They seek contact, but then resist by kicking, turning away, squirming or batting away offered toys. They continue to alternate between anger and clinging to the mother, and their exploratory play is inhibited (Holmes 1993, p. 105).

A later group was called insecure-disorganised:

4. Insecure-disorganised ('D') this small group…show a diverse range of confused behaviours including 'freezing', or stereotyped movements, when re-united with their parent (ibid).

This schema of different styles of attachment has informed numerous studies, particularly concerning the way in which childhood attachment experience influences adult attachment styles (Crittenden and Ainsworth 1989, p. 432). The primary importance of the infant-child relationship is repeated again by Winnicott, referring to 'primary maternal preoccupation': 'She knows what her baby is feeling like. No one else knows' (p. 15). Such a reading of maternal-infant bonds stands in stark contrast to some of the early modern historical accounts of motherhood. The importance of the mother was echoed by ethnologists and primate studies. In a speech delivered to American Psychological Association, Harry F. Harlow (1958, p. 673) wrote about the importance of mother-child bonding:

> From the developmental point of view, the general plan is quite clear: The initial love responses of the human being are those made by the infant to the mother or some mother surrogate. From this intimate attachment of the child to the mother, multiple learned and generalized affectional responses are formed.

Harlow, famed for experiments with maternal deprivation of infant macaques monkeys, viewed attachment as a necessary part of reproduction of the species, but added 'love' alongside the 'primary drives' of 'hunger, thirst, elimination, pain, and sex'.

> He does note that, for an infant, the mother is also the source of food, and that attachment through feeding facilitates the 'affectional bonds' of mother and infant.

The particular place of the mother was mixed with misogyny, and now, the mother, attributed the privileged status as special caregiver of infants, was seen as a threat in equal, if not more, regard. Mothering, the way that mothers related to their children, was increasingly seen as a site of fear and violence. Bettleheim (1967) regarded 'bad mothers' as responsible for children's autism. Winnicott wrote about the special need of the mother's ego to be of a certain strength, otherwise 'Where the mother's ego support is absent, or weak, or patchy, the infant cannot develop along personal lines, and development is then related, as I have said, more to a succession of reaction to environmental failure than to the internal urges or genetic factors' (Winnicott 1965, p. 17).

Bowlby had contributed to turning the mother figure into the site of all social ills. Anti-social behaviour of children, autism, learning difficulties, physical birth-defects, all now the result of the behaviour of the mother while the fetus was in the womb, or the particular style of mothering she engaged with the infant post-birth. Interpersonal relationships between mother and child were brought into hyper focus. The mother's way of breastfeeding, of weaning the child, of toilet training the infant, all these now became ways for psychiatrists to attribute psychological and personal psychopathy to the mother (Bettelheim 1967; Bowlby 1973; Rutter 1972). The psychiatrist Michael Rutter (Rutter 1972), looked at the way in which different forms of material deprivation—temporary, permanent, or intermittent—impacted on a child's development following divorce. The family unit, with the mother as the figure-head of caring for children, was continually empathized by these writers. Moreover, Bowlby demonstrated that children who were born into families with alcohol dependence, drug addiction or histories of violence and abuse did experience similar issues as adults and were less likely to enter into stable family relationships. The most contemporary research also confirms that drug and alcohol abuse, homelessness, poverty, violence and sexual abuse impact negatively on a child's ability to develop. The causes of autism or other psychopathological disorders were once conflated together as caused by bad mothering. In the 1980s, autism and general attachment studies parted company and autism became absorbed within the biomedical and neuroscientific model (Timimi and McCabe 2016a, b).

5.3 Children of the Wild

Machines are increasingly promoted as capable of providing relationality and care to human beings (Pino et al. 2015; Broadbent et al. 2009; Turkle 2012; Richardson 2015; de Graaf 2016), just as it has been argued that these can also be features of inter-species collaborations (Haraway 1991). Stories of children raised by wild animals are important in this regard. Narratives of feral children have long been a staple of myths, fairy tales and ways of imagining alternative ways of structuring society.

The feral child is a way of imagining what life would be like to live in the care of another species (Shattuck 1980; Sprehe 1961). Such stories include those about Kamala and Amala of Midnapore, southwest of Calcultta, reportedly raised by wolves[1]; Kaspar Hauser of Germany; and the Savage Girl of Songi in France. Accounts of feral children are controversial, politically

[1] It is reported this was a story fabricated by the missionary to receive funds, and suspicions were aroused because of racial arguments against 'native evidence' amidst the flourishing of Indian Independence (Newton 2002, pp. 192–193).

and racially charged narratives about the contrast between civilisation and savagery. They have often been used to justify the exploitation of colonial subjects, and animal cruelty and experimentation (Haraway 1991; Newton 2002). Most relevant for our purposes, however, is the question of whether animals are capable of raising children, and if so, what 'sort' of children would they raise?

Stories about children raised in the wild, by wolves, apes, pigs or dogs, are notoriously difficult to verify because these meetings took place when children were too young to recount the experiences. However, they do suggest what kind of relationship a child requires in order to develop a full range of human aptitudes. 'Feral children', when found, have been said to lack speech, language, motor skills and 'culture' (Candland 1995; Dennis 1941; Newton 2002; Shattuck 1994; Candland 1994). All children are totally dependent on caregivers for their very survival as they are born without speech, language, motor-control and any means of becoming self-sufficient by themselves. Thus, stories of feral children also connect with a deep fear of abandonment. Newton (2002), author of *Savage Girls and Wild Boys: A History of Feral Children* explains his fascination with these accounts:

> Every child fears abandonment. The displaced, unspoken anxieties of a family life fed those familiar, petty losses...and if a fear of abandonment partly informs this book, a sense of a faltering relationship with someone implacably remote, then here is also the guilty terror of abandoning. (p. 8)

What do these stories of children in the wild tell us? Zingg (1940) presents a chronological history of cultural accounts of feral children. These include those discussed by Linnaeus, who placed these accounts within a scientific study in 1758, but it was Buffon (1750) before him who introduced the concept of 'feral man' more widely in culture; and the French Enlightenment thinker Jean Jacques Rousseau in his *Discourse on the Origin of Inequality* (Rousseau 2000). Of 31 case studies of feral children, Zingg provides the ages of only four of the children described (p. 500). The youngest known 'feral child', the Leopard-boy of India, was abandoned at two years old and was rescued three years later. The scholars who recorded these cases were concerned about the behavioural, speech and language and cognitive differences among 'feral children'. For instance, it was reported that the Irish Sheep-boy would 'bleat' (ibid). Cases of 'feral' persons were divided into two categories: '(a) those who have wandered away into the wilds to survive by their own efforts unaided by human contact; and (b) children nurtured by wild or domesticated animals' (Zingg 1940, p. 493). The topic of feral children is shrouded in mystery, and mythical accounts merge with verified accounts—the latter originating from rural locations where it would be possible for an animal to encounter a human child. Together these accounts demonstrate that caregiving is just as important as meeting the child's physical needs, if not more. Affective bonds and socialisation are essential for a human child. What I am trying to emphasize here is that in reciprocal species-specific sociality, an *other* is already present and simultaneously engaged in human interactions, and that this other is not an exact copy, but a unique subjectivity and personhood. One human is not acting on another as an object, an 'I-It', but as an other, who is in turn developing and engaging their own species-specific sociality.

Sprehe (1961, pp. 161–167) put it thus:

> The data from feral man shows clearly the drastic effects that social deprivation can have on the human person. The absence of interaction with other human beings renders man *behaviorally* non-human. (my emphasis, p. 167)

That there are so few known cases of children surviving the wild demonstrates that caregiving relations with other humans are fundamental to a child's physical survival—a theme central to attachment theories of human intimacy (Bowlby 1997; Ainsworth et al. 1978). Moreover, these cases show us that the infant's capacities develop in relationship with others: speech, language, thought and motor skills are shaped by relational interactions, implying the relational caregivers have a profound impact on the development of infants and children (Crittenden and Ainsworth 1989).

We might ask what happens to our close kin, monkeys, who share some similarities with humans, if they are deprived of caregiving contact and nurturance with their own species? In *Affectional Responses in the Infant Monkey*—subheaded 'Orphaned baby monkeys develop a strong and persistent attachment to inanimate surrogate mothers' (1958, pp. 501–509)—Harlow and Zimmermann describe an experiment with 60 infant macaque monkeys who were separated from their mother 6 to 12 hours after birth and raised in a laboratory. Photographs and video-taped footage of these experiments show distressed monkeys, who appear to be in great suffering. Needless to say the experiments continued for the 'sake of science'!

Harlow and Zimmermann noted how the monkeys would cling to the cheesecloth blankets in their cages for comfort. The experiment was to test monkeys responses to a wire mother—a wire contraption with a food bottle attached—or a wire contraption with soft covering with no food bottle attached; these were collectively called 'mother surrogates' in the experiment. They write:

> Thus, both mothers provided the basic known requirements for adequate nursing, but the cloth mother provided an additional variable of contact comfort. That both of these surrogate mothers provided adequate nursing support is shown by the fact that the total ingestion of formula and the weight gain was normal for all infants fed on the surrogate mothers. (p. 423)

In these experiments, the monkeys were shown to favour the soft-covered wire 'mother' leading Harlow and Zimmermann to write, 'Although some of the infants reared by a wire mother did contact her, their behavior was similar to that observed in the home-cage fear tests' (p. 426). Harlow called the wire contraption the 'iron maiden'. The monkeys were observed in this condition to engage in 'convulsive jerking movements similar to the autistic behavior of deprived and institutionalized human children' (ibid).

The fundamental experience of infant-caregiver interaction and gestures are essential for I-you interrelatedness. According to Stawarska (2009), 'even though nonhuman primates engage in grooming and display affectionate care for one another, they do not seem to engage in reciprocal dialogic exchanges to the same extent human primates tend to across cultures' (p. 102).

Attachment theories and accounts of wild children do not diminish human relationships with animals. An animal might be a preferred *other* over a machine as nonhuman animals sustain themselves as living beings and are preoccupied with many of the same activities as humans: reproduction, eating, sleeping, defecating, rearing young. Harlow and Zimmerman's paper was intended to emphasise the importance of the mother-child bond, but instead, their paper offers confusing results. On the one hand, when the monkeys were deprived of species-specific contact, and given the 'choice' between a wire contraption and a soft-covered contraption (both of which are completely alien to the infant), monkeys preferred the soft-covered contraption: 'enough evidence is available to indicate that attachment formed to the cloth mother during the first 6 months of life is enduring and not easily forgotten' (p. 428). On the other hand, the monkeys were unable to mate or reproduce successfully and displayed distressed behaviours including stereotypical rocking back and forth. Surely it is this latter part that is truly important in this experiment. The survival strategies of the monkeys whilst in the labs and cages are conflated with the success of proving attachment theory.

What we find in all cases of feral children is a breakdown in human relations that give rise to these accounts. While Zingg quite rightly makes a distinction between cases of children of the wild and 'similar ones of children shut away from human association by cruel, criminal, or insane parents…They show the same effects from isolation' (1940, p. 493). In some cases, an animal may give a child (or adult) an important affective bond that is not provided by a parent or close kin. Take the case of Ivan Mishukov, raised in a family of violence and alcoholism, who, when he became homeless at around four years old, befriended wild dogs that became his substitute family. Ivan was abandoned most of all by those who were meant to make him feel safe—his family (Newton 2002). He survived because he was at an age where he could act with some independence from his 'carers'. Moreover, his case represents the exception rather than the rule: many children would perish in these circumstances.

5.4 The Machine as Surrogate Mother

In Bowlby's *Loss, Sadness and Depression* (1980), Volume 3 of his attachment series, he takes a sensitive look at the child's experience of sadness, mourning and loss. He describes four phases of mourning: 1. 'numbing' that lasts from a few hours to weeks, 2. 'yearning and searching' for the lost figure, 3. phase of 'disorganisation and despair', and 4. greater or lesser degree of 'reorganisation' (p. 85). This approach of taking seriously the lifeworlds of young children and describing what kinds of distress they can experience is integral to the idea of attachment theory. Though as a theory it has many critics, it at least opens up an empathetic exploration of the inner lives of children. The lives of infants and children have largely been ignored or neglected in Western culture, and children were primarily described as pre-logical (Piaget 2013, 1964) or unformed humans (Shorter 1750). It was not uncommon for thinkers to compare tribal peoples, children and neurotics in the same breath as all similarly engaged in magical thinking and were placed on a lower evolutionary ladder than adult Western male humans (Freud and Strachey 1950). There are a couple of exceptions to this in Western culture, found in the Old and New Testaments, including the tale of the Prophet, Moses; sent out by his mother Jachebed along the Nile to escape capture by the Egyptian Pharaoh who had ordered the killing of all first-born Israelites. Though we do not know much about Moses' early life, we do have a story of a man that begins with him as an infant. Even in Genesis, God created Man as an adult, not an infant, and woman was not even born of God, but the rib of Adam.

The most powerful Christian story of an infant is the story of Jesus. The New Testament tales describe his relationship with his mother, Mary, and his adopted father, the carpenter Joseph. The story of Jesus is perhaps one of the earliest narratives of an infant and child's life. The story of Jesus's birth was in a stable, visited by Three Kings:

> After they had heard the king, they went on their way, and the star they had seen in the east went ahead of them until it stopped over the place where the child was. When they saw the star, they were overjoyed.
>
> On coming to the house, they saw the child with his mother Mary, and they bowed down and worshiped him. Then they opened their treasures and presented him with gifts of gold and of incense and of myrrh. Matthew 2:1–12. (Bibles 2008)

Interest in children's lifeworlds and experience were largely ignored in literature up until the twentieth century which was developed by the new interest in childhood.

Human-robots' points of contact are incorrectly termed 'relationships': relationships, in the human sense of the term, indicate a reciprocal experience that is integral and crucial to the ongoing development of species-being. Is it possible to make analogies between human-robot points of contact and human-animal relationships, which are relationships with interpersonal interactions of various kinds, depending on the sociality of the animal? Animals may not speak for example, but they have communication systems that they use to signal their needs to humans, as pet owners can testify. However, for survival and development, any living being needs a member of a species specifically able to provide a plurality of attachment experiences.

Rather than seeing robots as an extension of living beings, they are more accurately viewed as tools—artefacts developed in certain ways. Tools can provide entertainment to humans and help to pass the time, particularly in societies with a breakdown in community associated with high levels of communal isolation. But such artefacts are not analogous to living beings.

To ascribe attachment theory to robotic machines (artefacts/sophisticated tools) is to ignore the need for species-specific sociality which is necessary for each species' ongoing existence.

Humans have a propensity to anthropomorphize objects, or to form 'attachments' to nonhumans, for example, the 'attachment' of a groom or bride to a wedding ring. In this case, the ring stands in for the attachment between the bride and groom, a symbol of partnership, commitment and co-experience. Artefacts that play this role in human culture have a significant symbolic meaning and are not easily comparable with other types of artefacts, such as cars or other consumer products, which are exchanged for others as a result of changing fashions or circumstances (e.g., a vehicle may be replaced because it would be more expensive to repair). Even companion animals can be seen as disposable: animal rescue centres are full of animals left behind by their owners. Yet, it is far rarer for humans to do the same to their own children.

Feminism has promoted a 'relational' rather than individualistic ontology (ontology refers to the study of the nature of being). Feminist philosopher Koehn writes: 'The fiction of a totally self-contained agent may be useful for some legal and political purposes. If we are not careful, however, we will come to think it a weakness to be unable to live a totally self-sufficient life devoid of emotional attachments' (1998).

Marketing companies report cases of adults who possess 'attachments' to their technological goods, including mobile phones, automobiles and computers (Grisaffe and Nguyen 2011; Vlachos et al. 2010; Roy et al. 2017).

Attachment as a study of interpersonal relationships, focusing on early caregiver infant/child-interactions, is not typically narrated through the lens of autism. Attachment theory, with its emphasis on mother-child interactions, is a controversial point of contact between the two disciplines. In Romanian orphanages, where extreme child-caregiver deprivation has occurred, there are higher rates of 'autism symptoms'. In addition, these children meet the 'descriptive criteria for autism' as do other children from other 'psychologically or socially vulnerable group[s]'. This, in turn, has led autism expert Michael Rutter to describe them as 'quasi-autistic' (Rutter et al. 1999 cited in Timimi and McCabe 2016a, p. 37).

The attachment formed between caregiver and child helps the child to make sense of themselves and others. Through caregiver-child interpersonal relationships, caregivers help infants to regulate feelings, soothe them and provide comfort, which all help the child develop affective, cognitive and motor functions. Research has demonstrated that in cases of typically-developing children, where there is no autism, autism-*like* traits can develop if the caregivers are threatening or abusive to the child.

There is the disturbing case of Elisabeth Fritzl. She was locked in a cellar and bore 4 children by her father, one of which was raised in the cellar. Elisabeth lived in the cellar for an astonishing twenty-four years as a captive, imprisoned and abused. There are more horrific stories of children treated as slaves, captives, punished and beaten—these children display severe difficulties, including social communication and attachment difficulties, cognitive and emotional problems, requiring life-long care and support in some cases. Alice Miller, author of *The Drama of the Gifted Child* (1987) explores how parents use their children to meet their own childhood needs which were not met by their own parents. Miller believes that using a child to meet adult needs prevents the child from experiencing their own needs. When compounded with physical violence and sexual abuse of children, this provides a basis for mental illness, anxiety and trauma. Interestingly, Baron-Cohen rejects the notion that childhood trauma can serve as the root for mental illness in his *Zero Degrees of Empathy* (2011).

Timimi and McCabe (2016a) also point out how orphaned and abused children display autism-*like* traits. These studies tell us that the way adults, particularly intimate caregivers, interact with their children does matter to how children develop. If robots are put in front of children as relational others, what will be the consequence for the way that children, with or without autism, develop? The attachment studies of Harlow (1956) demonstrate there are serious negative effects on infants if species-specific sociality and attachment is replaced by a surrogate machine. Even if robotic machines are created to look humanlike and move in humanlike ways, or speak, they still do not capture the full complexity of the interaction and meaning between caregiver and infant. Chloe Silverman (2011), in her book *Understanding Autism: Parents, Doctors, and the History of the Disorder*, explores how the absence of 'love' as a means to understand autism has put the medical professions at odds with families.

Silverman looks at the place of the autistic child in families and privileges the affective bonds between parents and their children. The affective love that is felt at the level of the family is not translated into the industrial-biomedical autism complex, but is ignored, bracketed off as irrelevant. Drawing on feminist models of the *ethics of care* (Sevenhuijsen 1998; Gilligan 1982), Silverman introduces 'love' into the conversation and reflects on how care is essential to human beings—both the provision and receipt of care.

5.5 Attachment to Products

Robots are not manufactured in a neutral economic system, but a capitalist economy that pours resources into new inventions and then markets these inventions by giving a purpose to them. Advertisements for products were called 'propaganda' by one the early pioneers of advertising and marketing, Edward Bernays (Bernays 2005[1928]).

Karl Marx, two centuries earlier, anticipated the rise of consumerism by referring to 'commodity fetishism' (pp. 76–77), writing:

> The mystical character of commodities does not originate, therefore, in their use-value...There is a definite social relation between men, that assumes, in their eyes, the fantastic form of a relation between things. (p. 77)

For Marx, the separation of product from producer created an illusionary sphere in which commodities had the potential to be more than their material contents. In such a way he predicted the growth of branding and advertising—strategies used by corporations to turn inanimate commodities into fetishes, objects with a life of their own. In his documentary, *A Century of the Self*, Adam Curtis describes the rise of consumer capitalism, shaped by the importance of the field of advertising and branding. Starting his story from the advertising campaigns of Edward Bernays, Curtis shows how the most valuable products were, at the beginning of the twentieth century, those that were durable and functional. However, the emphasis was quickly shifted to those that would appeal to the inner desires of the emerging American consumer class. Products were no longer merely inert artefacts (property) but supposed responses and solutions to intimate desires. Bernays was the nephew of Sigmund Freud and consulted his uncle on how to use psychoanalysis to promote corporate sales. Influenced by the ideas of Gustav Le Bon, who warned that crowds could be overtaken by irrational forces, Bernays wondered how to exploit the mass mind to sell products, writing:

> Propaganda assists in marketing new inventions. Propaganda, by repeatedly interpreting new scientific ideas and inventions to the public, has made the public more receptive. (Bernays and Miller 2005, p. 159)

In her book, *NO LOGO* (2010), Naomi Klein narrates a history of corporate branding, writing: 'Think of the brand as the core meaning of the modern corporation, and of the advertisement as one vehicle used to convey that meaning to the world' (p. 5). At the turn of the twentieth century, products were largely indistinguishable from one another, and branding became the way for corporations to personify their products. The purpose of branding these products was to show a humanised face to a population fearful of industrialisation.

The personification of products took off in the 1920s. An 'adman' named Bruce Barton was responsible for transforming the public persona of General Motors into a 'metaphor for the American family'. Barton said that the role of advertising was 'to help corporations find their soul' (ibid, p. 7). Adverts had to inform consumers about some new invention (as described by Bernays), and then convince them that their lives would be better if they used 'cars instead of wagons, telephones instead of mail and electric light instead of oil lamps' (ibid, p. 5).

If psychoanalysis helped to sell products in the early twentieth century, attachment theory was used to sell products in the late twentieth and twenty first centuries. Attachments to machines is one of the common frameworks for how technologists make a case for the utility and import of their machines (Grisaffe and Nguyen 2011, pp. 1052–1059; Vlachos et al. 2010, pp. 1478–1499; Huber et al. 2016, pp. 367–374).

Bibliography

Ainsworth, Mary D. Salter. 1978. The Bowlby-Ainsworth Attachment Theory. *Behavioral and Brain Sciences* 1 (03): 436–438.

Ainsworth, M.S., M.C. Blehar, E. Waters, and S.N. Wall. 1978. *Patterns of Attachment: A Psychological Study of the Strange Situation.* Hillsdale: Erlbaum.

Baron-Cohen, Simon. 2011. *Zero Degrees of Empathy: A New Theory of Human Cruelty.* Vol. 30. London: Penguin.

Bernays, Edward L. 2005 [1928]. *Propaganda. With an Introduction by Mark Crispin Miller.* Brooklyn: Ig Publishing.

Bernays, E. L., & Miller, M. C. (2005). *Propaganda.* Brooklyn, N.Y: Ig Publishing.

Bettelheim, Bruno. 1967. *Empty Fortress.* New York and London: Free Press; Collier-Macmillan.

Bibles, Collins Anglicised ESV. 2008. *The Holy Bible: English Standard Version.* London: HarperCollins.

Billington, T. 2016. Critical Autism and Critical Neuroscience: Towards a Science of Research and Practice. In *Re-thinking Autism: Diagnosis, Identity and Equality*, ed. K. Runswick-Cole, R. Mallett, and S. Timimi. London/ Philadelphia: Jessica Kingsley Publishers.

Bowlby, John. 1973. *Attachment and Loss: Anxiety and Anger*. Vol. 95. London: Hogarth Press.

———. 1980. *Attachment and Loss: Sadness and Depression*. Vol. 109. London: Penguin.

———. 1982. *Attachment and Loss*. Vol. 79. 2nd ed. London: Hogarth Press.

———. 2008. *Attachment*. New York: Basic Books.

Broadbent, Elizabeth, Rebecca Stafford, and Bruce MacDonald. 2009. Acceptance of Healthcare Robots for the Older Population: Review and Future Directions. *International Journal of Social Robotics* 1 (4): 319.

Candland, Douglas Keith. 1994. Introduction. In *The Forbidden Experiment: The Story of the Wild Boy of Aveyron. With a New Introduction by Douglas Keith Candland*, ed. Roger Shattuck. New York: Kodansha International.

Candland, Douglas K. 1995. *Feral Children and Clever Animals: Reflections on Human Nature*. New York: Oxford University Press on Demand.

Collins, G. 2016. Does a Diagnosis of ASD Help Us to Help a Person with Intellectual Disabilities? In *Re-thinking Autism: Diagnosis, Identity and Equality*, ed. K. Runswick-Cole, R. Mallett, and S. Timimi. London/Philadelphia: Jessica Kingsley Publishers.

Crittenden, P.M., and M.D. Ainsworth. 1989. Chapter 14 Child Maltreatment and Attachment Theory. In *Child Maltreatment: Theory and Research on the Causes and Consequences of Child Abuse and Neglect*, ed. Dante Cicchetti and Vicki Carlson, 432–463. London: Cambridge University Press.

Davis, K. 2016. How Rude? Autism as a Study in Ability. In *Re-thinking Autism: Diagnosis, Identity and Equality*, ed. K. Runswick-Cole, R. Mallett, and S. Timimi. London/Philadelphia: Jessica Kingsley Publishers.

de Graaf, M.M.A. 2016. An Ethical Evaluation of Human–Robot Relationships. *International Journal of Social Robotics* 8 (4): 589–598.

Dennis, Wayne. 1941. The Significance of Feral Man. *The American Journal of Psychology* 54 (3): 425–432.

Engels, Friedrich. 2010. *The Origin of the Family, Private Property and the State*. London: Penguin.

Freud, Sigmund, and James Strachey. 1950. *Totem and Taboo: Some Points of Agreement Between the Mental Lives of Savages and Neurotics*. London: Routledge and Kegan Paul.

Gilligan, Carol. 1982. *In a Different Voice*. Cambridge, MA: Harvard University Press.

Grisaffe, Douglas B., and Hieu P. Nguyen. 2011. Antecedents of Emotional Attachment to Brands. *Journal of Business Research* 64 (10): 1052–1059.

Haraway, Donna Jeanne. 1991. *Simians, Cyborgs and Women: The Reinvention of Nature*. London: Free Association Books.

Harlow, Harry F. 1958. The Nature of Love. *American Psychologist* 13 (12): 673.

Harlow, Harry F., and Robert R. Zimmermann. 1958. The Development of Affectional Responses in Infant Monkeys. *Proceedings of the American Philosophical Society* 102 (5): 501–509.

Hassall, R. 2016. Does Everyone with Autism Have the Same Underlying Condition? In *Re-thinking Autism: Diagnosis, Identity and Equality*, ed. K. Runswick-Cole, R. Mallett, and S. Timimi. London/Philadelphia: Jessica Kingsley Publishers.

Haydon-Laurelut, M. 2016. Critical Systemic Therapy: Autism Stories and Disabled People with Learning Difficulties. In *Re-thinking Autism: Diagnosis, Identity and Equality*, ed. K. Runswick-Cole, R. Mallett, and S. Timimi. London/Philadelphia: Jessica Kingsley Publishers.

Hodge, N. 2016. Schools Without Labels. In *Re-thinking Autism: Diagnosis, Identity and Equality*, ed. K. Runswick-Cole, R. Mallett, and S. Timimi. London/Philadelphia: Jessica Kingsley Publishers.

Holmes, Jeremy. 1993. *John Bowlby and Attachment Theory*. London: Routledge.

Huber, A., A. Weiss, and M. Rauhala. 2016. The Ethical Risk of Attachment: How to Identify, Investigate and Predict Potential Ethical Risks in the Development of Social Companion Robots. Conference Paper. HRI '16 The Eleventh ACM/IEEE International Conference on Human Robot Interaction, pp. 367–374.

Klein, Noami. 2001. *No Logo. Hammersmith*. London: Flamingo.

Koehn, Daryl. 1998. *Rethinking Feminist Ethics: Care, Trust and Empathy*. London: Routledge.

Latif, S. 2016. The Ethics and Consequences of Making Autism Spectrum Disorder. In *Re-thinking Autism: Diagnosis, Identity and Equality*, ed. K. Runswick-Cole, R. Mallett, and S. Timimi. London/Philadelphia: Jessica Kingsley Publishers.

Mallet, R., and K. Runswick-Cole. 2016. The Commodification of Autism: What's at Stake? In *Re-thinking Autism: Diagnosis, Identity and Equality*, ed. K. Runswick-Cole, R. Mallett, and S. Timimi. London/Philadelphia: Jessica Kingsley Publishers.

McGuire, A. 2016. Life Without Autism: A Cultural Logic of Violence. In *Re-thinking Autism: Diagnosis, Identity and Equality*, ed. K. Runswick-Cole, R. Mallett, and S. Timimi. London/Philadelphia: Jessica Kingsley Publishers.

Miller, Alice. 1987. *The Drama of Being a Child: And the Search for the True Self*. London: Virago.

Muskett, T. 2016. Examining Language and Communication in Autism Spectrum Disorder – In Context. In *Re-thinking Autism: Diagnosis, Identity and Equality*, ed. K. Runswick-Cole, R. Mallett, and S. Timimi. London/Philadelphia: Jessica Kingsley Publishers.

Newton, Michael. 2002. *Savage Girls and Wild Boys: A History of Feral Children*. London: Faber and Faber.

Piaget, Jean. 1964. Part I: Cognitive Development in Children: Piaget Development and Learning. *Journal of Research in Science Teaching* 2 (3): 176–186.

———. 2013. *The Construction of Reality in the Child.* Vol. 82. London: Routledge.

Pino, M., M. Boulay, F. Jouen, and A.-S. Rigaud. 2015. Are We Ready for Robots That Care for Us? Attitudes and Opinions of Older Adults Toward Socially Assistive Robots. *Frontiers in Aging Neuroscience* 7: 141.

Richardson, Kathleen. 2015. *An Anthropology of Robots and AI: Annihilation Anxiety and Machines.* Vol. 20. New York: Routledge.

Rousseau, Jean-Jacques. 2000. *Discourse on the Origin of Inequality.* Oxford: Oxford University Press.

Roy, Souvik, Abhilash Ponnam, and Santanu Mandal. 2017. Comprehending Technology Attachment in the Case of Smart Phone-Applications: An Empirical Study. *Journal of Electronic Commerce in Organizations (JECO)* 15 (1): 23–43.

Runswick-Cole, K., R. Mallett, and S. Timimi. 2016. Future Directions. In *Re-thinking Autism: Diagnosis, Identity and Equality*, ed. K. Runswick-Cole, R. Mallett, and S. Timimi. London/Philadelphia: Jessica Kingsley Publishers.

Russell, G. 2016. Early Diagnosis of Autism: Is Earlier Always Better? In *Re-thinking Autism: Diagnosis, Identity and Equality*, ed. K. Runswick-Cole, R. Mallett, and S. Timimi. London/Philadelphia: Jessica Kingsley Publishers.

Rutter, Michael. 1972. *Maternal Deprivation Reassessed.* Harmondsworth: Penguin.

Sayers, Janet. 1992. *Mothering Psychoanalysis: Helene Deutsch, Karen Horney, Anna Freud and Melanie Klein.* Harmondsworth: Penguin.

Sevenhuijsen, Selma. 1998. *Citizenship and the Ethics of Care: Feminist Considerations on Justice, Morality, and Politics.* London: Routledge.

Shattuck, Roger. 1980. *The Forbidden Experiment: The Story of the Wild Boy of Aveyron.* London: Secker and Warburg.

Shattuck, Roger. 1994. *The Forbidden Experiment: The Story of the Wild Boy of Aveyron. With a New Introduction by Douglas Keith Candland.* New York: Kodansha International.

Shorter, E. 1975. *The Making of the Modern Family*, 29–30. New York: Basic Books.

Silverman, Chloe. 2011. *Understanding Autism: Parents, Doctors, and the History of a Disorder.* Princeton: Princeton University Press.

Simon, G. 2016. Thinking Systems: 'Mind' as Relational Activity. In *Re-thinking Autism: Diagnosis, Identity and Equality*, ed. K. Runswick-Cole, R. Mallett, and S. Timimi. London/Philadelphia: Jessica Kingsley Publishers.

Sprehe, J. Timothy. 1961. Feral Man and the Social Animal. *The American Catholic Sociological Review* 22 (2): 161–167.

Stawarska, Beata. 2009. Between You and I: Dialogical Phenomenology.

Stone, L. 1979. *The Family, Sex and Marriage in England 1500–1800*. Vol. 43. Harmondsworth: Penguin.

Timimi, S., and B. McCabe. 2016a. What Have We Learned from the Science of Autism? In *Re-thinking Autism: Diagnosis, Identity and Equality*, ed. K. Runswick-Cole, R. Mallett, and S. Timimi. London/Philadelphia: Jessica Kingsley Publishers.

———. 2016b. Autism Screening and Diagnostic Tools. In *Re-thinking Autism: Diagnosis, Identity and Equality*, ed. K. Runswick-Cole, R. Mallett, and S. Timimi. London/Philadelphia: Jessica Kingsley Publishers.

Turkle, Sherry. 2012. *Alone Together: Why We Expect More from Technology and Less from Each Other*. New York: Basic Books.

Vlachos, Pavlos A., Aristeidis Theotokis, Katerina Pramatari, and Adam Vrechopoulos. 2010. Consumer-Retailer Emotional Attachment: Some Antecedents and the Moderating Role of Attachment Anxiety. *European Journal of Marketing* 44 (9/10): 1478–1499.

Winnicott, D.W. 1965. *The Family and Individual Development*. London: Tavistock.

Zingg, Robert M. 1940. Feral Man and Extreme Cases of Isolation. *The American Journal of Psychology* 53 (4): 487–517.

CHAPTER 6

Autism and the Machine

6.1 The Artificial Human

The importance of the machine for making sense of human experience and lived existence should not be underestimated. The machine *as* analogy, metaphor, physical artefact or social construct plays a powerful role in modern economies driven by science and technology. The power of the machine trope is in its ability to simplify the complexity of lived existence, hence rendering it more amenable to commercialisation—ideas about machines can be turned into commercial products. This chapter, then, is a study of the machine as used in robot and autism studies. The machine is used as a reference point to think about what it means to have autism, and vice versa. But what is this machine, and what are its characteristics? How does this machine map onto the complexity of human ontological experience? The machine model informs biomedical models of autism because autism is viewed like a system with parts. People with autism are believed to have an excess of parts—systemising; and a deficit of parts—empathising. This excess/deficit model informs biomedical narratives of autism, and it is the model picked up and made use of by robotics scientists, because machines, physically engineered ones, are also seen as a system of parts.

Machine models and artefacts are abstractions. However, some argue against this, suggesting that machine models constitute reality (Kurzweil 2000, 2005; Brooks 2002). Futurologists take the machine not as a representation of something, but as something it represents. This returns us

K. Richardson, *Challenging Sociality*, Social and Cultural Studies of Robots and AI, https://doi.org/10.1007/978-3-319-74754-5_6

to Baudrillard's *Simulacra and Simulation* (1994), whereby the map cannot be the territory, for the if the map of the territory and physical territory are the same, then the map would engulf the territory, and merge with it. So machines are reductive abstractions—something that can represent something other than what it is, while at the same time being enough of the thing represented for the abstraction to adequately capture something of that 'thing'. Machines are a way to make sense of complex phenomena, to see the workings of something as a system, and a system composed in parts. The parts of the system interact in unique ways and robotics scientists attempt to integrate the parts into a working whole. The machine and the Man-the-egocentric-individual are co-created; they belong together. The affective ties to kin, community, church and monarchy were replaced with an individual—a man who could live entirely for himself (Stone 1979; Hampson 1990). In order for Enlightenment Man to make sense of a chaotic universe, he rendered the complexity of the universe in the form he could manage—a machine. In the 1700s, clockwork and automatons were the first of these metaphors used (de Solla Price 1964, pp. 9–23), and a succession of machines have been used to signify the ages since. In our times we are more familiar with the personal computer, the internet (Castells 2000) and the cyborg (Warwick 2004; Haraway 1991). I want to suggest something else is at work apart from the attempt to reduce complexity with mechanical analogies. I want to suggest that the machine is the dominant analogy of the age because Enlightenment Man is primarily constructed as an egocentric, autonomous and separate individual; the machine is a mirror for egocentric Man.

6.2 Man as a Machine

The machine is quintessentially male; it represents a version of man that is not represented in woman. This association of the machine as man can be traced back in time to Julien Offray de La Mettrie's *L'homme Machine* (Man a Machine), written in 1747 (1912). The work is a critique of religion; it is a chaotic text, musing about the meaning of life in the 1700s. The text is punctuated with frequent references to Man as machine. For instance, 'Man is a machine—such a complex machine that it's initially impossible to get a clear idea of it or (therefore) to define it' (p. 3) and 'The human body is a machine that winds itself up, a living likeness of perpetual motion' (p. 5). The metaphor of the machine that de La Mettrie worked from was that of clockwork: 'A special feature of our machine is that every fibre in it, right down to the smallest, oscillates, and this natural oscillation is like that of a clock—it sometimes winds down' (p. 26). Clockwork as the metaphor for the machine was to the 1700s what the computer is to the twentieth century. The *cyborg* is arguably the figure of late twentieth and early twenty-first century (Haraway 1991). Donna Haraway turned the cyborg into a feminist symbol and used it as a political force to react against Enlightenment humanism: 'the cyborg has no origin story in the Western sense…An origin story in the "Western", humanist sense depends on the myth of original unity' (p. 292).

In the field of robotics and autism, the work of Simon Baron-Cohen has influenced a generation of robotics scientists who began to develop humanlike robots to help children with autism develop social interactive behaviours. Baron-Cohen's empathising-systemising theory proposes that there are essential differences between males and females relating to their 'hard-wired' preferences:

> A new theory claims that the female brain is predominantly hard-wired for empathy, and that the male brain is predominantly hard-wired for understanding and building systems. It is known as the empathising-systemising (E-S) theory.
>
> Empathising is the drive to identify another person's emotions and thoughts, and to respond to these with an appropriate emotion. The empathiser intuitively figures out how people are feeling, and how to treat people with care and sensitivity. (Baron-Cohen 2004, p. 2)
>
> Systemising is the drive to analyse and explore a system, to extract underlying rules that govern the behaviour of a system; and the drive to construct systems. (ibid, p. 3)

Baron-Cohen drew on an earlier idea, developed in the papers of Leo Kanner and Hans Asperger in the 1940s, about autistic people's interest in objects over persons (Frith 1991). Kanner proposed that the features of autism included 'extreme aloneness' (1943, p. 248) and a strong response to objects (ibid, p. 249). Asperger described how people typically live, interacting with their environment, but 'the autist is only himself (cf. the Greek word *autos*) and is not an active member of a greater organism' (ibid, p. 38).

Why is the inclination to 'systemise' not passed on from male 'father' to female 'daughter'? Why is neither parent able to shape their offspring of the opposite sex? According to Baron-Cohen's conceptualisation, males and females are closed and predetermined bodies, and only pass on the parts of themselves that are biologically determined. This is not so much a passing on, but an autonomous non-social replication, a mechanical reproduction of humans into male and female types. I have referred to this phenomenon as mechanical sociality (Richardson 2015), a machine that reproduces only itself. Stathern's *Gender of the Gift* (Strathern 1988) drew our attention to relations between women and men, not as outside each other, but interdependent with each other; women and men exist in interrelationship.

It was once believed that the absence of a space inside a man's body for the growing of a child is what reduces his capacity for caregiving and affectionate relationship, an argument explored and rejected in the work of Carol Gilligan (1982). In early attachment theories, the emphasis on the role of the mother at the expense of that of the father contributed to the stereotypical view that males lack emotional capacities. Gilligan has attempted to challenge this by showing how the affective life of young boys is suppressed, not by biology, but by adult intervention which encourages them to diminish and repress affective feelings. Girls, by contrast, are directed from a young age to put others before themselves and hide their intelligence. Girls actively make themselves less threatening to boys by diminishing their own independence and intellectual capacities. The socialisation of female and male children in this way distorts their spontaneous development as human beings, and brackets them into gender roles which serve to bolster male privilege. While early attachment theories often privileged the role of females in the caregiving story of infants, other ontological traditions that begin from I-you interrelatedness, such as Buber's *I and Thou* (1937) or dialogical philosophy (Stawarska 2009) see interpersonal relationship as a primordial ontological experience. These ontological perspectives emphasise privileged interrelatedness. Contemporary research in attachment has also taken into account the importance of caregiving provided by males and females in the attachment experience of human beings (Liss and Erchull 2012). Attachment does not end at childhood: studies also look at how attachment patterns influence the life-cycle of human beings from infancy to adulthood (Hazan and Shaver 1987, p. 511). Feminist discourses have impacted on theories by offering a 'relational' rather than individualistic ontology (ontology refers to the study of the nature of being). Feminist philosopher Koehn writes: 'The fiction of a totally self-contained agent may be useful for some legal and political purposes. If we are not careful, however, we will come to think it a weakness to be unable to live a totally self-sufficient life devoid of emotional attachments' (1998).

It is not the fact that men cannot produce children biologically or carry to term a child in their bodies that has prevented them from being fully involved in their children's lives. Instead, it has been the patriarchal social ordering which stigmatises fathers who participate in child-rearing, and maintains a gendered division of labour in the home (Liss and Erchull 2012; Sarkadi et al. 2008). In the Nordic countries of Sweden, Norway and Denmark, there is a strong political commitment to including fathers in children's lives, but this approach is not without difficulties as it means refocusing the traditional priorities of business and the state (Sarkadi et al. 2008).

Theories of systemising feed into stereotypes that men are not social and that they lack affective empathetic feelings. It is true that men are more likely to commit violent crimes against persons—a theme Baron-Cohen explores in *Zero Degrees of Empathy* (2011). However, there are complex social, economic and political explanations for the higher rate of male violence, to do with power, exploitation and privilege (Dworkin and MacKinnon 1988; Dworkin and Heiferman 1981; Gilligan 1982). There is ample evidence for the influence of environmental factors and childhood experience in precipitating men's violent behaviours. Yet, instead, many scientists resort to biological explanations (including testosterone, genes, brain differences) for the differing levels of violence between men and women. The work of Cordelia Fine in *Delusions of Gender* explores the myths of the biological basis of male violence, intelligence and emotion (2005). While there are sex differences between males and females, using genetic determinism to explain differences of behaviour between social groups echoes nineteenth-century eugenic ideas that justified inequalities against people of colour, women and those of a poorer socio-economic class (Thomson 1998). There are still millions of dollars of funding being poured into attempts to find a biological basis for autism despite these searches have so far proved futile (Grinker 2010; Eyal 2010).

There is strong evidence to suggest that autism is more common amongst males than females. As Timimi and McCabe argue (2016), there are gender disparities in certain kinds of diagnoses. Males, for instance, are more frequently diagnosed with Attention Deficit and Hyperactivity Disorder (ADHD) than girls. Some researchers claim that more females are on the autism spectrum than have been officially diagnosed. Girls on the spectrum are ignored because clinicians see autism as a male condition (Eyal 2010; Gould and Ashton-Smith 2011). It is not uncommon for some conditions to be distributed unevenly between the genders, but the reasons for this are open to debate. Many argue that essential differences between males and females account for this. Others, such as Moncrieff and Timimi (2010) look at the gendering of mental health conditions and the particularly harsh attitudes towards boys, particularly those from low-income families. Authority figures around these male children, Timimi, Gardiner and McCabe argue, are more like to see their behaviour as pathological.

Gender constructs about the roles of men and women are relevant here. According to Timimi and McCabe (2016), there are three dominant models of masculinity. First, the 'boys will be boys' model which emphasises heredity and a biological basis to male behaviour. Second, the environmental or 'boy code' model that is focused on the acculturation practices of becoming a boy and a man. Third is the model that tries to take into account the intersections between biological and environmental complexities. Yet, all the models serve to maintain the dominant understanding of how males differ from females: the males' lack of empathy, social withdrawal and isolation. These stereotypes are not without impact: evidence suggests that from the moment infants are born in European and North American countries, the traits and behaviours, and even the physical environment of male and female infants are all gendered (e.g., a girl's bedroom is more likely to have pink featured somewhere). This in itself contributes to gender differentiation (Fine 2005, pp. 202–212).

6.3 Autism as a Machine-State

If man is the quintessential machine, then the condition of autism is the machine par excellence, stripped of empathy, biologically programmed to systematise. Analogies are commonly drawn between men and machines, and autistic persons and machines. The focus on autism as a non-social state, combined with a preference for things makes it the condition par excellence of the *digital age*. The luminaries of the digital age, including Steve Jobs at Apple, Mark Zuckerberg at Facebook and Bill Gates at Microsoft, are all believed to be on the autism spectrum. Baron-Cohen has proposed that autistic people are well-represented in the STEM fields: Science, Technology, Engineering and Mathematics (Baron-Cohen et al. 1997, 2009, 1998). These are also fields in which females are underrepresented.

For someone like Baron-Cohen, this underrepresentation has little to do with the exclusion of women from education and public life over centuries, misogyny, patriarchy, or violence that women typically experienced, because he believes that the field of opportunity has since been levelled—particularly as women can vote (awarded in 1928 for women aged 21 and over), enter universities (though before 1920 women could not matriculate), and enter politics (in 1918 was the first female elected representative of a UK parliament, but as an Irish Republican she did not take up her seat). In 2017, we do not have a 50–50 balance of the sexes in parliament, with only 195 females out of 650 members. In the United States women hold 19.6% of the seats in Congress (out of 535 in total), with only 21% of seats in the United States Senate and 19.3% in the United States House of Representatives.[1]

Baron-Cohen is cited in a BBC news article arguing for a biological basis to mathematical ability. He argued that in over 100 years of the Fields Medal (the equivalent of the Nobel Prize in maths), there have been no female winners. The reason given by Baron-Cohen for this is foetal testosterone (Fine 2011, p. 99). Bioscience has become a way for elitist ideas about class, race and sex to be recast as 'neutral' and 'scientific'. Elitism refers to a critique levelled against those who believe they are superior to, and worthier than, other human beings. Goodley (2016) suggests that an elitist ethnoclass is created based on economic, political, and able-bodied exclusion, in which the concept of white Man is overrepresented in ontological body of what is understood to be human. Up until the age of anti-racism, feminism, working class resistance, and anti-colonialism, the human was synonymous with property owning man. Goodley starts from the position that 'people associated with the label of autism are fundamentally members of the human species and, simultaneously, the presence of autism demands that we think again about what it means to be a member of humankind (p. 146).

It is important to know what concept of *the human* is being utilised by researchers in robotics and biomedicine. Is a man their primary reference point when they think about what constitutes the human, or do their conclusions apply equally to woman *the whole spectrum of humanity*? Historically, male rationality and perspective taking has been preeminent in how the human has been construed, white European and North American privileged men also ignored other men of colour and men of lower economic and social status. European and North American 'Man' was then a subset of men. For much of human history, in various different ways, women were excluded from political, economic, legal and academic

[1]Women in the U.S Congress 2017. Rutgers Eagleton Institute of Politics. [Online] Accessed 3.11. 2017 <http://www.cawp.rutgers.edu/women-us-congress-2017>

affairs. Humans with mental or physical disabilities also framed as inferior. This hierarchy of status, based on power operates in the world today, despite ongoing efforts to democratise political, legal, economic and educational institutions (Sinclair 1962; Richardson 2016b; Dworkin and MacKinnon 1988; Margaret and Naffine 2001).

This presumption that autism and male supremacy in STEM subjects are rooted in biology, rather than culture, reproduces old hierarchies newly legitimated in the language of science and technology. The consequences of a distorted understanding of the human in robotics and biosciences impacts on the real lived experiences of human beings that are labelled 'autistic'. For instance, DSM clinical guides rely on a set of ideas about what autism is, and how it can be recognised:

> B. Restricted, repetitive patterns of behavior, interests, or activities, as manifested by at least two of the following, currently or by history (examples are illustrative, not exhaustive; see text):
>
> 1. Stereotyped or repetitive motor movements, use of objects, or speech (e.g., simple motor stereotypies, lining up toys or flipping objects, echolalia, idiosyncratic phrases).
> 2. Insistence on sameness, inflexible adherence to routines, or ritualized patterns or verbal nonverbal behavior (e.g., extreme distress at small changes, difficulties with transitions, rigid thinking patterns, greeting rituals, need to take same route or eat food every day).
> 3. Highly restricted, fixated interests that are abnormal in intensity or focus (e.g., strong attachment to or preoccupation with unusual objects, excessively circumscribed or perseverative interest). (Diagnostic Criteria, Autism Spectrum Disorder 299.00 (F84.0). (American Psychiatric Association 2013)

What does the description above tell us about autism? The description tells us nothing about the child's potential, or openness or creativity or spontaneity, but frames autism, and the person, through a lens of reduction. This psychological approach to reading a human being, as if they are diagnostic criteria, is problematic. It reduces the complexity of what it means to be human.

Goodley (2016, p. 147) goes further in his challenge to the biomedical establishment: 'Others have suggested that autism is nothing more than a myth perpetuated by the biopolitics of psychiatry and psychology; disciplines obsessed with their own sense of entitlement and legacy' (p. 147). Does autism exist? Does schizophrenia or borderline personality disorders? How about the frequency with which diagnoses can change from one category to another? When this happens, did a person have one condition, and then it transformed into another? Are all these categories ultimately based on a reductive idea that reduces humans to formulas in texts and clinical practices, rather than living beings?

Representational forms of the machine indicate a particular kind of elitist egocentric Man: signified as objective, authoritative, non-emotional, detached and systemising. If the machine becomes another way of speaking about Man, then what is the non-machine? If systemising Man, expressed in its truest form is autism, then what is Woman? Is she the antithesis of the machine? In contemporary robotics, the gendering of robots plays an important role, as their socially interactive qualities are simultaneously ascribed a gendered form. It is common to apply a male or female appearance to human-made artefacts and this is quite common in the design of dolls, statues, effigies, puppets or automata (Wood 2002). There are differences emerging in the design of male- and female-looking robots. The female inspired robots fall into two main categories—service robots and sex robots (Richardson 2016a). This gendering of robots into males and females (gender is a social construction, sex is biological) is also relevant to those robots developed to help children with autism. Robots do not have sexed bodies as they are not alive and do not possess a species-specific living sociality, but they can be contrived to have a 'gender' and a form to present as either male or female. Gender roles are assigned on the basis of economic, political social orderings (Jeffreys 2014). Females are presented as submissive and unintelligent so males can feel powerful and intelligent over and above them (Dworkin and Heiferman 1981).

Bio-medical diagnoses and services cannot easily be disentangled from the person with autism, as clinical and medical services are present in the early developmental stages of the child's life. The scenario frequently runs thus: Parents wonder about the physical, social or cognitive differences of their child, they seek medical guidance, help and support. The medical practitioners process the child's state in the vocabulary of the medical profession, the parents take on the medical language as a way to make meaning of the child's situation. The child is exposed to medical interventions where the language is repeated, alongside physiotherapeutic, educational or pharmaceutical interventions.

The child grows into an adult with a history of particular kinds of medical interventions and medical language is thus woven into the person's story and self-understanding (Brunwick-Cole 2016). Milton (2016) proposes that autistic people favour 'sociologically situated model[s] of autism' (p. 794). Baggs alters this language, transforming the concepts of the medical machine that operates around her, and offering something different. Yet disentangling from the bio-medical language is fraught with difficulties, as this language is the more widely circulated and informs popular understanding.

Milton (2016) observes that autistic contributions to conceptions of autism have been the least heeded. Associations with particular 'brands' of autism, such as systematising—the type that technologists like to suggest they have, and parents like to highlight in their children—are what Goodley calls 'ableist'—defined by Wolbring as showing 'the favoritism for certain abilities- for example, cognition, competitiveness or consumerism- and the often negative sentiment towards the lack of favoured abilities' (cited in Goodley 2016, p. 155).

Fine (2010) examines the theory of 'brain sex' proposed by Baron-Cohen (2004) that male/female brains are 'hard-wired' with systemising/empathising qualities by looking at responses to statements such as '*if there was a problem with the electrical wiring in my home, I'd be able to fix it myself*' or '*when I read the newspaper I am drawn to tables of information*'. Choose these and you are systemising. Conversely, if you chose '*I can easily tell if someone else wants to enter a conversation*' or '*I really enjoy caring for other people*', you would be an 'empathizer' (p. 15).

As Fine points out, 'Asking people to report on their own social sensitivity is a bit like testing mathematical ability with questions like "*I can easily solve differential equations*", or assessing people's motor skills by asking people to agree or disagree with statements like "*I can pick up new sports very quickly*"' (p. 16). Moreover, such descriptions feed into stereotyped understandings of gender roles. There is no way of ascertaining whether the person who responded positively to the statement '*if there was a problem with the electrical wiring in my home, I'd be able to fix it myself*' correctly represented their abilities. How they answer may thus itself be shaped by cultural expectations of their gender. Asking participants to identify their gender beforehand, as this experiment did, thus acts as a prime, subconsciously informing the participants that their gender is relevant to what is being 'tested'.

Asperger's view of women was expressed openly in his paper describing a mother whose behaviour resembled that of her son: he was surprised as he expected women to be 'more emotion than intellect' (p. 41).

Goodley remarks when referring to autism, the extreme male-brain and its originator, Baron-Cohen, who is Professor of Developmental Psychopathology and a Fellow at Trinity College, Cambridge: 'I am sure the irony is not lost on you that autism's linkage with extreme maleness is being developed in an elite British university that might be conceived as being at the heart of the British empire that historically helped spawn ethnoclass man across the globe' (2016, p. 155).

6.4 Robot as Analogy to Autism

In the field of technology, analogies between people with autism and machines are frequent. There is a field called Human-Robot Interaction (HRI) that is dedicated to exploring interactions between humans and machines. In a 2013 doctoral dissertation, entitled *Robots for social skills therapy in autism: evidence and designs toward clinical utility*, Kim (2013) starts from this assumption concerning people with autism: 'Given the high prevalence of interest in machines and devices found among children with ASD, the use of robots as a highly motivating object is the fundamental motivation of HRI research autism interventions' (p. 185). Likewise, the authors of the paper *Human-Robot Design for Robot-Assisted Intervention for Children with Autism based on E-S Theory* write: 'Male brains belong to Type S [systemizing] for the most part, reflected in the their outstanding performances in mathematics, engineering, etc.. While females have more Type E [empathizing] brains, reflected in their sensitivity in affections and feelings. E-S theory indicates that autistic individuals have an Extreme Type S [extreme –systemising] cognitive feature, which results in intact or even superior systemizing with impaired empathizing' (Li et al. 2016).

The rationale for robot therapies is drawn from ideas that people with autism are primarily non-empathetic, and not socially organised like typically-developing human beings. It is widely believed that males (and autistic people) are non-empathic and thereby more attracted to machines in the form of robots or computers, or non-social processes such as STEM subjects.

Social studies of science and technology challenge the view that science or technologies are practices that exist outside the social realm (Hasse 2002; Keller 2009; Helmreich 1998). Emily Martin's research on how language about male and female sex cells is used in science demonstrates that such language is not detached and dispassionate, but figuratively speaks of the heteronormative conquest of female 'damsel' eggs by 'chivalrous' male sperm (1991). Helmreich's study of evolutionary computer modelling in Palo Alto revealed that computer scientists employ biologically-based arguments to inform their computer programs.

Computer programmes are inorganic artefacts: they are not sexed, do not reproduce or engage in the living species-specific iterations. Computer scientists might apply evolutionary and biologically inspired models to computer programs, but these programs exist as commercially produced artefacts, not living systems. Science and technology do not exist autonomously outside of social and cultural existence.

Autism has become a way to think about machines and vice versa.

Computer scientist Gal A. Kaminka authored a paper entitled: *Curing Robot Autism: A Challenge* (2013). If the title is surprising, the content is more so: 'Almost all robots are autistic; very few humans are' (2013, p. 801). He continues: 'The analogy is to treatment of high-functioning autism by behavioral therapy, whereby the patient (in our case, the robot) is given rules to follow blindly to improve social functioning, without real internalization of their rationale' (p. 801). I have been told innumerable times by robotics scientists that *children with autism have a special affinity to machines*, and by parents of children with autism that the reason their child likes to play with an iPad is because they are autistic.

A focus on imitation in psychotherapeutic interventions and through robot therapies is identified; as autistic children show difficulties in imitation behaviours. A study by Nadel et al., (2004) proposed to show how robot systems could be designed by using autistic children as the comparison; they write, 'Adopting a bottom-up perspective, we investigated how infants and low-functioning children with autism develop low-level imitations, while we explored in parallel which implementations are needed for autonomous robots to develop the same kind of low-level imitations' (p. 47). Again the autistic person is the 'platform' of comparison with the autonomous robot. Imitation is arguably important for learning and human cognitive, emotion and motor development but why do robotics scientists believe it is possible to transfer the learning and imitation difficulties of autistic children into machines?

If it was once dehumanising to compare people with disabilities to machines, in the cyborg age it is commonplace. We are told We're All Cyborgs Now! As the materialistic worldview of sci-tech inspired capitalism takes hold, so the way in which bodies and persons are conceived is through a materialistic 'corporate' prism. Nevertheless, part of the motivation for comparing people with disabilities and mental health difficulties to machines in robotic science research is that these researchers are drawing on hierarchies of ability: people with physical and intellectual disabilities are seen as closer to machines. The able-bodied human being is the target goal of many researchers, but they must get to this able-bodied "Man" (Richardson 2010). This is not to say that there are not other artificial intelligence and robotics projects which are lead directly by the fantasy of reproducing *elite* Man in the machines (Kurzweil 2000, 2005; Brooks 2002).

Davies (2016) would refer to this as 'ableism'—'a network of beliefs, processes and practices that produces a particular kind of self and body (the corporal standard) that is projected as the perfect species-typical and therefore essential and fully human. Disability is cast as a diminished state of being human' (Campbell 2001, p. 44 cited in Davies 2016, p. 134).

People with disabilities and mental health problems are deemed unwell and therefore in need of services, therapy or care, which, robotics scientists believe, can be replaced by robots. People with disabilities generally have less power than those working in robotics—a high status profession—which means that how they conceptualise autism (as a robotic state) gains traction in wider culture. The machine-model of autism works for a state of non-sociality, but sociality is intrinsic to the human condition, not an optional extra. The machine model informs a script to make sense of men and people with autism, but the machine has been formed to exclude the primary interrelatedness that is integral to all human beings' existence.

The etymological origins of the term 'machine' is from the ancient Greek mēkhos, meaning 'contrivance', which means (1). Something contrived esp. a mechanical device or a plan (2). An act of contriving, esp. deceitfully and (3). An inventive capacity (OED 1995, p. 291).

Autism Speaks' is an autism advocacy organisation based in the United States and sponsors research on autism science and education. In 2014 they launched a high profile media campaign to promote MSSNG a genome sequencing program. The MSSNG campaign or the E-S brain theory, or the DSM criteria, machines are always reductions based on the foreclosure of I-you interrelatedness. Instead the machine is synonymous with autism because the link between them is egocentric Man.

Promising studies that involve adults with autism and Asperger's dispel the myths of autism as a non-social or non-empathetic sociality. Anthropologists Och and Solomon (2010, pp. 69–92) coined the term 'autistic sociality' to open up the conversation about autism as a social state, rather than an asocial one. New studies led by mothers of autistic children (Runswick-Cole 2016), autistic academics (Milton 2014), and activists such as Amanda Baggs and the Critical Autism Network

(Runswick-Cole et al. 2016) are leading the way in challenging the powerful authority of the biomedical industry. Collins (2016) argues the diagnostic schema wastes considerable resources by organising resources around clinical criteria that are irrelevant to living intellectually disabled person and their actual needs.

Studies with adults with Asperger's and their parents reveal that autistic adults are grappling with social reality and interpersonal relationships (Robinson 2014). Milton (2014), an academic and autistic person had this to say about these connections:

> The "machine-like" metaphor…often used to describe autistic cognition and knowledge acquisition is critiqued…autistic people (across the spectrum) are indeed social beings, albeit perhaps a more idiosyncratic or outsider social experience and expressions of social agency. (p. 795)

The transference from socially inspired robots to children with autism is predicated on an analogical comparison, according to which, something within one domain can be transferred to another; that which is transferred is something that is equal to both. The analogous state of the child with autism and the social robot is something conferred by the bio-medical frameworks around each. The consequences of the mechanical representation of bioscience of autism have now fed into the development of robotics. Such representations are neither neutral nor commercially innocent.

Bibliography

American Psychiatric Association. 2013. *Diagnostic and Statistical Manual of Mental Disorders (DSM-5®)*. American Psychiatric Pub.

Baron-Cohen, Simon. 2004. *The Essential Difference*. London: Penguin.

———. 2011. *Zero Degrees of Empathy: A New Theory of Human Cruelty*. Vol. 30. London: Penguin.

Baron-Cohen, Simon, Sally Wheelwright, Carol Stott, Patrick Bolton, and Ian Goodyer. 1997. Is There a Link Between Engineering and Autism? *Autism-London* 1: 101–109.

Baron-Cohen, Simon, Patrick Bolton, Sally Wheelwright, Victoria Scahill, Liz Short, Genevieve Mead, and Alex Smith. 1998. Autism Occurs More Often in Families of Physicists, Engineers, and Mathematicians. *Autism* 2 (3): 296–301.

Baron-Cohen, Simon, Emma Ashwin, Chris Ashwin, Teresa Tavassoli, and Bhismadev Chakrabarti. 2009. Talent in Autism: Hyper-Systemizing, Hyper-Attention to Detail and Sensory Hypersensitivity. *Philosophical Transactions: Biological Sciences* 364 (1522): 1377–1383.

Baudrillard, Jean. 1994. *Simulacra and Simulation*. University of Michigan Press.
Brooks, Rodney L. 2002. *Flesh and Machines: How Robots Will Change Us*. London: Allen Lane.
Campbell, F.A. 2001. Inciting Legal Fictions-Disability's Date with Ontology and the Abieist Body of the Law. *Griffith L. Rev* 10: 42–62.
Castells, M. 2000. The Rise of the Network Society: Economy, Society and Culture v.1: The Information Age: Economy, Society and Culture Vol 1. Oxford UK and Malden MA: Wiley-Blackwell.
Collins, Graham. 2016. Does a Diagnosis of ASD Help a Person with Intellectual Disabilities? In Re-thinking Autism: Diagnosis, Identity and Equality, ed. K. Runswick-Cole, R. Mallett, and S. Timimi. London/Philadelphia: Jessica Kingsley Publishers. Pp. 204–220.
Davies, Kim. 2016. How Rude? Autism as a Study in Ability. In *Re-thinking Autism: Diagnosis, Identity and Equality*, ed. K. Runswick-Cole, R. Mallett, and S. Timimi, 132–145. London/Philadelphia: Jessica Kingsley Publishers.
Davies, Margaret and Ngaire Naffine. 2001. Are Persons Property? Legal debates about property and personality. Dartmouth, Ashgate. Pp. 1–202.
de La Mettrie, Julien Offray, and I.I. Frederick. 1912. *Man a Machine*. La Salle: Open Court Publishing Company.
de Solla Price, Derek J. 1964. Automata and the Origins of Mechanism and Mechanistic Philosophy. *Technology and Culture* 5 (1): 9–23.
Dworkin, Andrea, and Marvin Heiferman. 1981. *Pornography: Men Possessing Women*. Vol. 1. London: Women's Press.
Dworkin, Andrea, and Catharine A. MacKinnon. 1988. *Pornography and Civil Rights: A New Day for Women's Equality*. Organizing Against Pornography.
Eyal, Gil. 2010. *The Autism Matrix*. Cambridge: Polity.
Fine, Cordelia. 2005. *Delusions of Gender: The Real Science Behind Sex Differences*. New York: Icon Books Ltd.
———. 2010. *Delusions of Gender: The Real Science Behind Sex Differences*. London: Icon Books.
———. 2017. *Testosterone Rex: Unmaking the Myths of Our Gendered Minds*. London: Icon Books.
Frith, Uta. 1991. *Autism and Asperger Syndrome*. Cambridge: Cambridge University Press.
Gilligan, Carol. 1982. *In a Different Voice*. Cambridge, MA: Harvard University Press.
Goodley, D. 2016. Autism and the Human. In *Re-thinking Autism: Diagnosis, Identity and Equality*, ed. K. Runswick-Cole, R. Mallett, and S. Timimi, 146–158. London/Philadelphia: Jessica Kingsley Publishers.
Gould, Judith, and Jacqui Ashton-Smith. 2011. Missed Diagnosis or Misdiagnosis? Girls and Women on the Autism Spectrum. *Good Autism Practice (GAP)* 12 (1): 34–41.

Grinker, Roy Richard. 2010. Commentary: On Being Autistic, and Social. *Ethos* 38 (1): 172–178.
Hampson, Norman. 1990. *The Enlightenment*. Vol. 4. London: Penguin.
Haraway, Donna Jeanne. 1991. *Simians, Cyborgs and Women: The Reinvention of Nature*. London: Free Association Books.
Hasse, Cathrine. 2002. Gender Diversity in Play with Physics: The Problem of Premises for Participation in Activities. *Mind, Culture, and Activity* 9 (4): 250–269.
Hazan, Cindy, and Phillip Shaver. 1987. Romantic Love Conceptualized as an Attachment Process. *Journal of Personality and Social Psychology* 52 (3): 511.
Helmreich, Stefan. 1998. *Silicon Second Nature: Culturing Artificial Life in a Digital World, Updated with a New Preface*. University of California Press.
Jeffreys, S. 2014. *Gender Hurts: A Feminist Analysis of the Politics of Transgenderism*. London/New York: Routledge.
Kaminka, G.A. 2013, May. Curing Robot Autism: A Challenge. In *Proceedings of the 2013 International Conference on Autonomous Agents and Multi-agent Systems*, 801–804. International Foundation for Autonomous Agents and Multiagent Systems.
Kanner, L. 1943. Autistic Disturbances of Affective Contact. *Nervous Child* 2 (3): 217–250.
Keller, Evelyn Fox. 2009. *The Century of the Gene*. Cambridge: Harvard University Press.
Kim, Elizabeth Seon-Wha. 2013. *Robots for Social Skills Therapy in Autism: Evidence and Designs Toward Clinical Utility*. Yale University.
Koehn, Daryl. 1998. *Rethinking Feminist Ethics: Care, Trust and Empathy*. Psychology Press.
Kurzweil, Ray. 2000. *The Age of Spiritual Machines: When Computers Exceed Human Intelligence*. New York: Penguin.
———. 2005. *The Singularity Is Near: When Humans Transcend Biology*. Penguin.
Li, Chaochao, Qingxuan Jia, and Yongli Feng. 2016. Human-Robot Interaction Design for Robot-Assisted Intervention for Children with Autism Based on ES Theory. *IEEE*.
Liss, Miriam, and Mindy J. Erchull. 2012. Feminism and Attachment Parenting: Attitudes, Stereotypes, and Misperceptions. *Sex Roles* 67 (3–4): 131–142.
Mallett, Rebecca, and Katherine Runswick-Cole. 2016. The Commodification of Autism: What's at Stake? In *Re-thinking Autism: Diagnosis, Identity and Equality*, ed. K. Runswick-Cole, R. Mallett, and S. Timimi, 110–131. London/Philadelphia: Jessica Kingsley Publishers.
Martin, Emily. 1991. The Egg and the Sperm: How Science Has Constructed a Romance Based on Stereotypical Male-Female Roles. *Signs: Journal of Women in Culture and Society* 16 (3): 485–501.

Milton, D.E. 2014. Autistic Expertise: A Critical Reflection on the Production of Knowledge in Autism Studies. *Autism* 18 (7): 794–802.

Moncrieff, J., and S. Timimi. 2010. Is ADHD a Valid Diagnosis in Adults? *No. BMJ: British Medical Journal* (Online), 340.

Nadel, Jacqueline, Arnaud Revel, Pierre Andry, and Philippe Gaussier. 2004. Toward Communication: First Imitations in Infants, Low-Functioning Children with Autism and Robots. *Interaction Studies* 5 (1): 45–74.

Ochs, Elinor, and Olga Solomon. 2010. Autistic Sociality. *Ethos* 38 (1): 69–92.

Oxford English Dictionary (OED). 1995. *The Concise Oxford Dictionary of Current English*. First edited by H.W. Fowler and F.G. Fowler. Ninth Edition. Edited by Della Thompson. Oxford: Clarendon Press.

Richardson, Kathleen. 2010. Disabling as Mimesis and Alterity: Making Humanoid Robots at the Massachusetts Institute of Technology. *Etnofoor* 22 (1): 75–90.

———. 2015. *An Anthropology of Robots and AI: Annihilation Anxiety and Machines*. Vol. 20. New York: Routledge.

———. 2016a. Sex Robot Matters: Slavery, the Prostituted, and the Rights of Machines. *IEEE Technology and Society Magazine* 35 (2): 46–53.

———. 2016b. Are Sex Robots as Bad as Killing Robots? *IoS*.

Robinson, J. 2014. *Participatory Research with Adults with Asperger's Syndrome: Using Spatial Analysis to Explore How They Make Sense of Their Experience*. Doctoral Dissertation, De Montfort University, Leicester.

Runswick-Cole, K. 2016. Understanding This Thing Called Autism. In *Re-thinking Autism: Diagnosis, Identity and Equality*, ed. K. Runswick-Cole, R. Mallett, and S. Timimi, 19–29. London/Philadelphia: Jessica Kingsley Publishers.

Runswick-Cole, K., R. Mallett, and S. Timimi. 2016. Introduction. In *Re-thinking Autism: Diagnosis, Identity and Equality*, ed. K. Runswick-Cole, R. Mallett, and S. Timimi, 7–15. London/Philadelphia: Jessica Kingsley Publishers.

Sarkadi, Anna, Robert Kristiansson, Frank Oberklaid, and Sven Bremberg. 2008. Fathers' Involvement and Children's Developmental Outcomes: A Systematic Review of Longitudinal Studies. *Acta Paediatrica* 97 (2): 153–158.

Sinclair, Thomas Alan. 1962. *Aristotle the Politics*. Harmondsworth: Penguin Books.

Stawarska, Beata. 2009. *Between You and I: Dialogical Phenomenology*. Athens: Ohio University Press.

Stone, L. 1979. *The Family, Sex and Marriage in England 1500–1800*. Vol. 43. Harmondsworth: Penguin.

Strathern, Marilyn. 1988. *The Gender of the Gift: Problems with Women and Problems with Society in Melanesia*. Vol. 6. University of California Press.

Thomson, Mathew. 1998. *The Problem of Mental Deficiency: Eugenics, Democracy, and Social Policy in Britain C. 1870–1959*. Oxford University Press.

Timimi, S., and B. McCabe. 2016. What Have We Learned About the Science of Autism? In *Re-thinking Autism: Diagnosis, Identity and Equality*, ed. K. Runswick-Cole, R. Mallett, and S. Timimi, 30–48. London/Philadelphia: Jessica Kingsley Publishers.
Warwick, Kevin. 2004. *I, Cyborg*. Urbana: University of Illinois Press.
Wood, Gaby. 2002. *Living Dolls: A Magical History of the Quest for Mechanical Life*. London: Faber & Faber.

CHAPTER 7

Critical Autism Studies and Robot Therapy

7.1 A New Horizon

Critical autism studies (CAS) is a new, exciting field of study and practice that is critical of bio-medical perspectives of autism. Led by a group of pioneering academics, activists and adults with autism, critical autism studies proposes a move away from the biomedical and deficit/excess model of autism that is so prevalent in psychiatry and promoted through the psychological sciences and clinical practice, towards a more accepting framework for autism. The manifesto of this new approach was launched in an edited volume: *Re-thinking autism: diagnosis, identity and equality* (Runswick-Cole et al. 2016). The authors make several proposals:

1. To challenge the evidence base for biomedical models of autism.
2. To explore the impact of the diagnosis on the lived experience of people so labelled.
3. To offer a conceptual critique of the production, consumption, and commodification of autism in the twenty-first century in a global context.
4. To explore ideas for service provision and practice that move beyond a focus on diagnosis.

There are multiple lines of criticism of the contemporary biomedical industry of autism, and the identity politics of neurodiversity and disability

K. Richardson, *Challenging Sociality*, Social and Cultural Studies of Robots and AI, https://doi.org/10.1007/978-3-319-74754-5_7

that are part of autism culture. The huge investment of resources into finding a biological basis to autism, combined with widely differential figures of who has autism, Critical Autism Studies is needed more than ever to bring clarity to a multimillion-dollar industry. As Runswick-Cole et al. (2016) argue, 'despite the vast amount of resources spent on finding a genetic basis to autism, there is none forthcoming. And rather than accept that the premises in the biomedical model might be flawed, researchers argue that the need is for more resources – this perpetuates a cycle of investment and funding for the sake of a restrictive biomedical model of autism that seemingly does not exist. The authors of the volume continue: 'The more failures that pile up, the more 'complex' autism genetics must be according to advocates, whilst the most likely explanation for this finding – that there is no such things as genes that 'cause' autism – remains unmentionable'(ibid, p. 36).

CAS are informed by critical anti-psychiatry, advocacy and social models of disability and difference methodologies. Latif (2016) writes: 'Psychiatry continues to be perceived as unscientific, and keeps getting associated with violence, stigma and untreatability…The medical model, which has a narrow focus on within-person experiences and the biology of the individual, creates a decontexualised and simplistic idea of the problem, disengaging us from our social responsibility as a society, and enhancing our dependence on doctors or "experts", rather than trying to use our own problem-solving skills. The alternative of a values-based biopsychosocial model embraces broader conceptualisations of the origins of complex behaviour' (p. 296).

Foucault's anti-psychiatry perspective (1967) examined the way in which madness and psychiatry were invented to serve an *othering* function. Early medicine treated madness as deviant from the norm, and took over from the role of religion. While religious institutions used terms like 'demonic possession', medicine and psychiatry appropriated the 'possessed' into their analysis, diagnosis and treatment and called this 'madness' and 'insanity'.

Madness was contrasted with rationality and Reason, the throne of the Enlightenment. Scottish psychiatrist R. D. Laing led the way by exploring the interpersonal dynamics of the nuclear enclosed family, and how family structures and interpersonal relationships led to mental health issues. Laing opened up new ways of thinking about mental health in a hostile environment of practitioners reluctant to take on board new perspectives or examine the premises and assumptions of their psychiatric practices.

Today, the *British Medical Association* (BMA) and *The Wellcome Trust*, two important biomedical institutions, encourage new scientists to consider humanistic approaches to medicine and psychiatry and actively support funding and training in these areas. However, more needs to be done. The dominant theories that grip autism studies are also the economically lucrative ones. If change is to occur at all levels, biopsychosocial and attachment theories, rather than bio-science exclusively, need to be more widely used to make sense of human beings.

The Critical Autism Studies perspective is not without its problems. Two chapters that are part of the collected volume use similar kinds of mechanical 'system' led thinking that is endemic in robot-autism therapy. Simon (2016) and Haydon-Laurelut (2016) proposed a 'systemic' thinking schema to make sense of the complexities of human beings. There are some useful insights from the chapters that draw upon this theorising, but they are mapped onto a 'systemic' structure that promotes reductionist approaches to the ontological nature of being human. For example, Simon (2016) explores this in her chapter *Thinking Systems 'Mind' As Relational Activity*, writing, 'as professionals, we need to shift our skills from diagnostic activities to relational curiosity to develop collaborative ways of conversing with people who have been diagnoses with Asperger's Syndrome and others in their networks. Through immersing ourselves in dialogue with people and their social and professional networks, we can bring forth evidence of people with Asperger Syndrome having unique and useful expertise, which can play a part in resolving concerns and improving communication between those in their networks' (p. 269). This is sound advice, but can be done without reference to what Simon calls 'cybernetic systems': 'within these cybernetic systems are feedback loops, mechanisms for managing cyclical or unusual events, rule-bound and rule-creating' (p. 275). Sigh! The power of reductionist and system-based thinking is so powerful that it is everywhere, even in radical approaches to autism. Like systemising-empathising or E-S brains, these abstractions do not exist in real people but in cognitive schemas.

This systemic schema is reproduced by Haydon-Laurelut (2016), who again uses the narratives of autistic people, opening up different kinds of dialogical possibilities that were foreclosed through clinical models of rightness or wrongness of communication, but again defaults to formulaic descriptions: 'systemic therapists work with systems' (p. 227). While systems are a good way to think about machines, they are problematic when applied to I-you interrelatedness between people. Moreover, a system implies an observer who can make sense of the system, rather than a process of plurality, inconsistencies, and open-endedness.

Simon draws on the language of cybernetic systems e.g., systems, networks. The dialogical element of their proposal for working with children with autism is in keeping with the Buber tradition of dialogue, however.

7.2 Labels and Language of Autism

Most, if not all of the literature on robot therapy for autism rarely accounts for the changing meaning of autism over time. Autism as a category is not fixed in time and space, and its diagnosis and relevance to medicine and society is constantly shifting. What constitutes autism shifts over time, but so do the ways in which the clinical category is applied to people. The subjectivity of the medical professional also impacts on who and who does not get a diagnosis of autism, even though diagnostic instruments have been developed to have some formal 'objective' measure (Latif 2016, p. 292). Intelligence quotient (IQ) was one such measure for example, and an IQ of around 80 was typical until the category of autism was expanded to include Asperger's Syndrome in the early 1990s. With the 1994 version of DSM-III, clinicians began to include persons with Asperger's, who typically have a higher IQ, into the category of autism. Some categories of the DSM have been taken out (e.g., homosexuality in 1973), others transformed (e.g., autism) and new ones added (e.g., attention deficit hyperactivity disorder (ADHD) (see Latif 2016).

Grinker (2008) describes how labelling a child with complex needs as 'autistic' helped the family to get much needed federal resources in the United States. There are financial incentives for parents of children with intellectual disabilities to gain a diagnosis of autism compared to other disabilities. The 'deficit model' of autism is challenged by disability and difference advocates and new empirical studies (Grinker 2008, 2010; Ochs and Solomon 2010; Brownlow and O'Dell 2009). The CAS network has moved the boundaries of debate to examine the assumptions that inform models of autism (Runswick-Cole et al. 2016; Runswick-Cole 2016; Timimi and McCabe 2016a, b; Hassall 2016, Ortega and Rios 2016; McGuire 2016; Mallett and Runswick-Cole 2016; Davis 2016; Goodley 2016; Hodge 2016; Collins 2016; Haydon-Laurelut 2016; Rosqvist et al. 2013; Milton 2014; Evans 2013).

Russell (2016) has problematised the notion of diagnosing autism at an early stage of a child's development, arguing that the behaviours that prompt such a diagnosis may be susceptible to change as the child grows up. She refers to studies of Self Fulfilling Prophecies (SFP) and looked at sociological experiments of self-belief. She cites the 'Pygmalion' study by Rosenthal and Jacobson (1966) where 'researchers, posed as educational psychologists, went to a school and tested a class of pupils. They then told the teacher that 20% of these pupils were "intellectual bloomers": potentially brighter than the others, despite these 20% being selected quite randomly. When the children were retested a year later the "bloomer" pupils really did do better in intelligence tests' (p. 261).

Russell explores 'Labelling Theory', which suggests that labels guide people's interactions with others. I can confess to my own experience of this in the early days of my research, as I looked at the children through the 'descriptions' of autism. I asked myself questions like: Did the child look at me? Was the child more interested in the toys than me? I used a biomedical lens as a frame to make sense of the children rather than experiencing the child as they presented themselves to me in that moment. I did not see the children as I-thou in the Buber sense of being open to what was happening between us. Rather I wondered about the children through the lens of clinical and academic descriptions of autism, I acted towards the children as I-It (Buber 1937).

Russell invites us to consider the labelling lens and how it interrupts the formation of knowledge through interpersonal interaction. Approaching a child as a developmental being, rather than an individual with a condition, and providing healthcare and educational support for them accordingly, would give space to the child's subjectivity. Rather, a child and their families could be provided for, in a way, that remained attentive to the child's capabilities, open to the child's potential to change over time, and conscious of the reality that a child is always a person, not a label, a category, or a shorthand that is convenient for an industry. In therapeutic approaches, reducing distance between 'expert' and 'patient' has led to innovations in healthcare and given over more agency to a person with a difference or disability to manage this themselves (Rogers 1997). The autism advocacy movement is a strong and important voice for adults with autism, and the social disability model allows adults with autism to become the 'experts' of their own lived existence.

While the idea that autism is really a sign of genius may be comforting for those diagnosed with autism, this nevertheless divides and categorises as 'It': systemisers, geniuses, idiot savants, low-functioning or high functioning, obsessive-compulsive, antisocial or non-empathetic.

The systemisers category enacts violence towards the person with intellectual disabilities and differences called 'autism'. It allows others to look upon people with autism as lacking in sociality and empathy. A systemiser and empathiser does not exist anywhere other than a text, in much the same way that a map does not reproduce its territory. While the map is an abstraction of the territory, humans cannot be reduced to abstractions. Except human beings are reduced to abstractions through clinical diagnoses. The robot as a therapeutic treatment for this othering is a logical follow-on.

Parents are faced with considerable dilemmas when they are trying to make sense of their child who is developing differently or has different kinds of social needs and behaviours. It is vital that a challenge to biomedical models of autism does not become a reason used by state governments, who already are pressed for resources and promote austerity strategically, to exploit these arguments by the Critical Autism Studies network to eliminate funding for services to the intellectually disabled and different populations. Russell (2016) explains:

> Parents faced considerable dilemmas when confronted with the prospect of an autism diagnosis. Although diagnosis may lead to services, respite and a partial explanation for a child's behaviour, parents also considered the potential devaluation, stigma and rejection that may result. Diagnosis *may* lead to services, but often does not, thus often operates as 'promise' that may or may not be fulfilled…The autism diagnosis therefore represents a double-edged sword for many parents'. (original emphasis p. 262)

The use of particular kinds of language can influence what sorts of treatment, medical or technological, are offered. One unsuccessful campaign was launched by US charity Autism Speaks in 2014, titled 'MSSNG'. The MSSNG campaign referred to a genome sequencing project, but individuals with autism took issue with the explicit 'neuro-typical' language in the public launch. This led to a general backlash from the autism community, particularly adults with Asperger's and parents of children with autism. Also, there are some adults with autism that reject any biomedical approach that aims to 'cure' autism.

Autism advocates see autism as part of their identity. Bagatell (2010), for example, describes attending an Asperger's group with a member wearing a T-shirt that read 'eye contact is overrated' as group members sought to subvert normative assumptions about what is socially normal. In fact, norms around eye-contact vary by culture: in some cultures, it is considered disrespectful for a young person to maintain eye-contract with an older person or a female person to maintain eye-contact with a male (McCarthy et al. 2006). Thus, considering an autistic person's discomfort with eye-contact to be problematic is culturally, rather than biologically, informed.

7.3 It and Things

Historian of Science, Ian Hacking has written extensively about the way in which categories of persons come about through psychiatric and clinical diagnoses. He led the *Making Up People* project exploring how psychiatric categories fused with identity politics.[1] Identities are organised through medical categories: autism, ADHD, personality disorders, all gave rise to new identities articulated through biomedical categories. Hacking sets the scene for autistic biography:

> We now watch, hear and read a great deal about what it is like to be autistic—in autobiography, biography and fiction. Autism narrative is a new genre: not expert reports by clinicians or reflections by theorists, but stories about people with autism, told by the people themselves, or their families, or by novelists, or by writers of stories for children.

Runswick-Cole (2016) and Milton (2014) have discussed autistic identity as a placeholder of difference, often a reaction to the othering language of those institutions that want to 'cure' autism. The 'expert' led narratives of autism can come into conflict with the autism advocacy movement and their allies, as mentioned above in reference to the 2015 Autism Speaks MSSNG (missing) campaign to detect the autism genes. The description of the MSSNG project is thus:

> MSSNG is a groundbreaking collaboration between Autism Speaks, Google and the research community to create the world's largest genomic database on autism. MSSNG's goal is to provide the best resources to enable the identification of many subtypes of autism, which may lead to better diagnostics, as well as personalized and more accurate treatments[2]

[1] Hacking, I. 2006. Making Up People. London Review of Books, Vol. 28, No 16. 17th August 2005, pp. 23–26. [Online] Accessed 19.8.17 < https://www.lrb.co.uk/v28/n16/ian-hacking/making-up-people>

[2] Autism Speaks "MSSNG" Accessed 25.8.17 < http://research.mss.ng/>

This led to a huge backlash from autistic people who reacted against the idea that there was something 'missing' in them. One tweeter, Autistic Vegan wrote:

> '@AutismSpeaksPBC @google I am #ActuallyAutistic and I am not #**MSSNG**! Stop silencing us. We are right here, and valuable as we are'.
>
> Whether it is the 'expert' or the 'identity politics' of the person with autism, autism is primarily framed through bioscience and the provision of health in state economies. It was psychiatry and the psychological sciences who created autism.

In contrast with biomedical models of autism are ones that place emphasis on interpersonal encounter and interrelatedness. One such proponent of this approach is Martin Buber in his work *I and Thou* (1937). For Buber it is through dialogue, presence and meeting, that 'they' become a 'you' (or for Buber a mysterious thou). For Buber, when 'I' is spoken, 'you' or 'it' is also implied. So the 'I' in whatever form it takes is important. If the I does not engage with the you as it is, there is a discrepancy, a mismatch between the actual person, and the concept in the head that is interacted with, and hence distortions occur. If the you is merely an instrument to be used—an 'it'—then the I who is opposite the it experiences themselves as an it. If two I-its meet, then they itify each other. If the I-it is reproduced more widely in society, then I-It become the orthodoxy.

Latif (2016) argues for a 'value-based approach'. With regards to a 'narrow reductivist' approach, she writes: 'Human behaviour involves social and moral dimensions, and cannot be understood without taking into consideration the values of that individual, their family and indeed their broader cultural milieu' (p. 289). She continues that attempts to relate to a person with a developmental difficulty without this approach will produce 'social control through the medium of politically sanctioned state institutions' (p. ibid). Beata Starwarska in *Between You and I: Dialogical Phenomenology* (2009) writes about the inseparability of the I and You, rejecting the egocentric (individualist) tradition in classical Western philosophy to propose an alternative 'polycentric view' (p. 11).

Disability activists such as Mel Baggs are determined to speak of their experiences in their own language.

> My name is Mel Baggs – you might know me as Amanda, but I've started going by my middle name again. I am a disabled writer and artist. I am

> cognitively disabled, physically disabled, chronically ill, developmentally disabled, and psychiatrically disabled, so I have experience with a lot of different disability communities.

I often feel like an outsider in all of them, but the one I feel most at home in is the developmental disability self-advocacy community. This is because it's not a community based on a diagnosis, but a community based on a history of shared experiences, and I have lived in the developmental disability service system for pretty much my entire adult lifetime. I was born in 1980, so I am 33 years old at the time of writing this (Amanda Baggs, Accessed 25.7.2017[3]).

The autism industry is a multibillion-dollar industry. The condition of autism is also presented as a large scale financial problem for welfare state provision of medical, educational or other kinds of assisted living costs. In Australia, autism is a multibillion-dollar industry and research estimates the costs of autism to society at 4.5–7 billion Australian dollars (Horlin et al. 2014). In the UK, autism costs at least £32 billion per year.[4] These costs are incurred by therapeutic services and interventions, medical care, special education, support services for the family, respite care, pharmaceuticals, and home healthcare. In addition to the state costs there is also an abundance of alternative therapies and autistic services and products, including autistic biographies (Hacking 2009). Even the way in which autism is 'sold' to the public is by presenting it in certain ways.

In the egocentric tradition with its 'I' without a 'you', which underlies the capitalist system, products become the substitute 'you'. Theories built on the foundations of the individualistic 'egocentric tradition', such as the biological basis to autism, or psychoanalytic traditions that inform attachment theory, are easily eviscerated by corporate organisations as they employ them in order to sell products.

Latif (2016) notes that mental health provision, services, treatment, diagnosis and support are economic systems. Pharmaceutical companies rely on 'expert' diagnoses for the development of new lines of drugs and treatment. 'These powerful institutions use the "expert" model of doctors

[3] https://ballastexistenz.wordpress.com/about-2/

[4] Autism is the most costly medical condition in the UK. [Online] Accessed 19.8.17 <http://www.lse.ac.uk/website-archive/newsAndMedia/newsArchives/2014/06/Autism.aspx>

as a mechanism for profit-making through the labelling of individuals' (p. 294). The new inventions are what produce income and valorisation (the ability to grow value).

While technology is often seen as 'value for money' compared to the arts, a report found that 25% of IT projects fail immediately and up to 50% require major rework. Since the scale of failures in IT does not prevent investment in these areas, it might be useful to ask what function does the field of information technology play in the reproduction of capitalism? Though arguably large gains can be made with technologies, such as computers, the economic input into these areas by funding bodies of nation states (such as DARPA or the European Union) mean that the number of failures or instances of poor returns on investments are hidden from the public.

Runswick-Cole (2016) is an academic and a mother of an autistic child. She describes, in a powerfully moving way, the multiple models of autism she has encountered in making sense of her son's experience and how the label of autism has served as a powerful and prefigured marker. Unlike many autobiographical accounts of autism, which pick up and reproduce the narratives of autism—'stimming', 'systemitising', 'neurotypical'—instead Runswick-Cole offers something completely unique; she writes:

> Given that I am arguing that the category of autism is of limited use to those diagnosed with it, you may be surprised to hear that, like other 'autism moms', I also believe autism stole my child. But my son has not been taken by the autism, the biological disorder; I do not believe that he is locked within an autistic world, out of reach and alone. Rather I argue the *thing* called autism that has stolen by child is a contemporary cultural construct, a story. This autism tells of an all-encompassing label, a narrative lens that accounts for every aspect of my son's behaviour and personality, so much so that there is nothing left of him that cannot be accounted for without reference to his autism. It is *this* autism that has stolen by child. (p. 24)

Runswick-Cole's story resonated as true with me, as I have heard this 'contemporary cultural construct' expressed through the same script repeated again and again, almost word for word by parents of children with autism, by autism educators, by clinical experts, and routinely and frequently by roboticists who develop robots for children with autism. The language of systematising, theory of mind, and imitation persistently circulate in the culture of autism, unquestioned.

In my role as an ethics advisor to a European project I have tried to introduce other perspectives beyond the biomedical model. To date, these have had little effect—in fact, the way in which ideas about autism are picked up in wider culture has even become part of the discourse used to 'explain' differences between men and women. In August 2017, newspapers reported that a Google employee had written a memo rejecting 'diversity' within the company and arguing that natural differences between males and females accounted for variance in technical aptitude. He invoked the language of Baron-Cohen's systematising/empathising theory, almost reproducing it verbatim. He wrote that women, on average, have more:

> Openness directed towards feelings and aesthetics rather than ideas. Women generally also have a stronger interest in people rather than things, relative to men (also interpreted as empathizing vs. systematising).

These two differences in part explain why women relatively prefer jobs in social or artistic areas. More men may like coding because it requires systemising and even within SWEs, comparatively more women work on front end, which deals with both people and aesthetics.[5]

In technology circles, systemising/empathising theory fits neatly with a sense of male superiority. It has also become a framework for making arguments for why *robots* should be developed for children with autism. Runswick-Cole continues: 'This thing called autism has such a grip on my child and my family that attempts to steal him back simply become evidence of my pathology' (Brunswick-Cole 2016, p. 25). The widespread introduction of robots into the therapeutic context is another example of 'therapeutic violence' (Timimi et al. 2010 cited in Runswick-Cole, 2016 p. ibid, p. 26) where the needs of the powerful bio-medical and therapeutic research community, their particular ideals, goals and perspectives provide the dominant framework in which autism is understood, treated and challenged.

Robotics has now become part of that community. The researchers who use robots are firmly situated within the biomedical model of autism and the systemising/empathising theory of autism.

[5] Conger, Kate. Exclusive: Here's the Full 10-Page Anti-Diversity Screed Circulating Internally at Google. Gizmodo [Online] 8/125/17 Accessed 12.8.17 <http://gizmodo.com/exclusive-heres-the-full-10-page-anti-diversity-screed-1797564320>

Brunswick-Cole observes that a school in the United States is still using electric-shock behavioural therapy. The power of Brunswick-Cole's perspective is in inviting us to stop thinking about identifying a child's autism and assigning labels to the child or her or his behaviour, but to be with the person, to widen our range of acceptability. Her words make a lot of sense.

Just prior to my fieldwork in the school, I learned about autism through books about it. When I arrived at the school, I also looked for 'signs' of autism. Did the children look at me? Did the child give me eye-contact? Did the child prefer interacting with the toys, or robot or crane more than with me? I had not wholly looked upon those children in the moment or tried to understand them as individuals, as urged by Brunswick-Cole. I looked at the children through the texts I had read about them. As Brunswick-Cole suggests, this lens gets in the way of meeting and knowing her son, who is more than the texts. While there are efforts to move away from labelling, the fact that this book has 'autism' in the title, and I refer to autism and robots, means that I am part of the narrative of autism. Yet it is a narrative based on an illusion of the bio-medical system.

7.4 Robot Therapy

There is now a well-established field of robotics focused on research exploring the therapeutic benefits robots for children with autism (Esteban et al. 2016; Coeckelbergh et al. 2016b; Pop et al. 2013; Peca et al. 2014; Hashim and Yussof 2017; Scassellati; Simut et al. 2012; Huijnen et al. 2016). The potential efficacy of this robot therapy is explained as thus:

> The clinical use of interactive robots is a promising development in light of research showing that individuals with ASD: (a) exhibit strengths in understanding the physical (object-related) world and relative weaknesses in understanding the social world, (b) are more responsive to feedback, even social feedback, when administered via technology rather than a human, and (c) are more intrinsically interested in treatment when it involves electronic or robotic components. (Diehl et al. 2012, p. 250)

This research has proposed that robots can help children with imitation skills (Nadel et al. 2004), acting as a predictable entity for social interaction (David et al. 2014; Tapus et al. 2012), in eliciting emotional expressions and behaviours from autistic children (Kozima et al. 2005), and acting as a way to engage children in story-telling games (Simut et al. 2012). Robots have also been studied as diagnostic tools (Scassellati

2007). The primary purpose for some of the robot studies was to play games with the autistic children, usually imitation games with the robot (Robins et al. 2005; Dautenhahn and Werry 2004; Dautenhahn 2007).

In the study of the effectiveness of robots for helping children develop social behaviours, responses to the robot in the study were mixed. Many children responded differently from the way the children interacted with the toy crane. When children came to see the robot, they often sat at a table on a chair and faced the robot directly. The remote-controlled crane was on the floor. The robot had a body that was humanoid and so the children sometimes poked the robot's 'face' and 'eyes'.

In experiments investigating how children with autism interact with robots, the environment is structured in a particular way: the robot is often seated or standing on table-tops, while a chair is often provided for the child. Moreover, the affordance of a humanlike robot, or a vaguely anthropomorphic robot like Keepon, impacts how users can interact with it. A robot that has a bodily form and humanlike face offers different kinds of affordances.

In the enclosed and sparse setting of the 'experiment' room, the robot and crane were visibly distinct. While children did interact with the robot and the crane according to the affordances they offered, in the context of the environment, this did not happen on many occasions.

One child, May, enjoyed taking off the robot's wig which revealed wires and circuit boards underneath. Another child, Noah, enjoyed slapping the robot across its face and poking the robot directly in its eyes. Mark and Patrick imitated the remote-controlled actions of the robot. Other children, such as John and Ben preferred to imitate the buzzing sounds of the crane's motors as they took charge and operated the remote-control. Anthony did not spend time at all interacting with the robot, instead he would roam around the room, ignoring my attempts to encourage him to play: our time together rarely lasted more than around three minutes.

Research has also been conducted on attitudes towards the use of robots in child healthcare, with concerns about human replacement being outweighed by the urgency for therapy (Costescu and David 2014, p. 3; Coeckelbergh et al. 2016a, pp. 47–65). This means that parents are so desperate for any kind of intervention and support to help them they will be open to new technologies. The European funded DREAM (Development of Robot-enhanced therapies for children with autism) studies that developed robots to administer Applied Behavioural Analysis therapies, using a single-case study experiment had in 2016 found the robot had not improved the child's social behaviour as demonstrated by improvements in (1) Turn-taking, (2) Imitation or (3) Joint attention. Research studies (Scassellati et al. 2012, pp. 275–294) have been difficult to reproduce and the findings in many studies report on the secondary characteristics (how many times does the child engage in imitation or joint attention) rather than rigorous clinical studies. The DREAM research is not yet completed but reports on the progress of the project can be found here: http://www.dream2020.eu/deliverables/. The DREAM consortium team have a strong commitment to disclosing the research findings to the public and encouraging public dialogue.

Such claims about the effectiveness of robots and technology have to be situated within a larger context of biomedical and neuroscientific approaches to making sense of autism. Let us put robots in the context of these other therapeutic artefacts. Certain studies report on the therapeutic benefits of puppets for children with autism (Trimingham 2010, pp. 251–265), but such studies are drowned out by the current compulsion for all things technological. Animals, puppets and robots can impact on the learning experiences of all young children, neurotypical or not. But when the agent is seen as closer in kind to the child with autism, the tool has the potential to increase the child's interest and engagement with everyday tasks such as brushing teeth or going to bed. This was the case, for example, in Trimingham's (2010) studies of the 'humble puppet' (p. 263). This is significant since puppets can be controlled by a parent or care-giver and are far more affordable and easy to operate than a robotic platform. Moreover, the relationship between adult and child is privileged in puppet therapy for children with autism. 'Puppets, when operated by someone who establishes a rapport with the child, and particularly when their use is sustained over time, are dependable quasi-'transitional objects' that offer a 'break' from feeling so out of joint with the world (p. 262). The results from studies with puppets shows real potential, yet there is less interest shown in them than in robotic alternatives.

Dogs, and other animals have similarly shown positive results for helping autistic children feel calmer and facilitating their communication and interactions (Solomon 2012, pp. 109–126) Dogs have been close attachment figures for humans and have a long history of human-dog companionship lasting thousands of years (Haraway 2003). Temple Grandin, an author and professor of animal science at Colorado State University with autism, has written extensively about the connections between the lifeworlds of animals and the lifeworlds of autistic people. Grandin has emphasised the affinities between them, particularly with regard to their stimulus-fear response of fight or flight. Grandin and children and adults with autism report experiencing anxiety and panic at a profound level as a result of their physiological response to stress situations. During my time at the school for children with autism, I would notice the children, huddled up in the corner with a blanket or jacket over their heads during the lunch break. Owing to the Victorian style of the building, the high ceilings and large windows let in excessive light, from which the children took cover in any way available. These self-soothing actions reveal both the children engaging in self-care in stressful situations, and also seeking to communicate their feelings to adult observers.

7.5 Recognition

Recognition refers to the ability to see something. According to the Oxford English Dictionary it comes from the transitive verb 'to recognise' meaning 'identify (a person or a thing) as already known; know again; 2. To realize and discover the nature of; 3. Realise or admit; 4. Acknowledge the existence, validity, character or claims; 5. Show appreciation of, or reward; 6. Treat or acknowledge; 7. Allow (a person) to speak in a debate (1995, p. 1147).

These ways of framing recognition imply one individual giving way to another, acknowledging them. Recognition is not neutral; cultural conventions, knowledge systems and practices shape how the other is seen. In the political sciences, Recognition is a social justice-inspired narrative informed by anti-colonialism, the othering of peoples through discourses, practices and regimes of violence. Goodley (2016) draws on the way in which 'Othering' is a form of dehumanisation, allowing those with more power to determine *what* others with less power are. This is a theme explored by Michael Taussig (1993) in *Mimesis and Alterity: A Particular History of the Senses.* Taussig studies how colonial representation of the Cuna indigenous peoples was disconnected from the real lived experiences of the peoples: the Cuna and other indigenous peoples in South America met by colonialists served as a mirror for the colonialists, projecting back to them their own repressed ideas about themselves. The colonialists could not recognise the indigenous peoples as independent, and self-willed. Instead the colonialists read the actions of the indigenous people through their own narrow lens of *their* perspective.

Taussig invites us to think about the way in which traditional artefacts can be taken from their location in Panama into museums in Stockholm without any transformation in the idea of what it means to be 'European'. However, when indigenous peoples take artefacts produced in Europe back into their culture, it is argued that they are losing their identity as indigenous people and becoming *Westernized*. Much like the colonial encounters between colonialists and indigenous peoples, the egocentric tradition excludes I-you interrelatedness.

If the myth of Echo and Narcissus has any place in European and North American cultures of domination, it can be traced from the time of slavery, through to colonialism, to contemporary patriarchy. There is not *more* 'narcissism' or *less* 'narcissism' as political sociologist Christopher Lasch (1991) argued. Rather, the egocentric perspective of Western elites is endemic to a political-economic system that forecloses equality. While political systems may, in theory, guarantee rights for all their citizens, economic inequalities are permitted to exist and are seen as distinct from political equality. This economic inequality persists and worsens because of the ability to see certain humans as tools, meriting less pay and fewer benefits than their employers.

Bibliography

Bagatell, Nancy. 2010. From Cure to Community: Transforming Notions of Autism. *Ethos* 38 (1): 33–55.

Bertilsdotter Rosqvist, H., C. Brownlow, and L. O'Dell. 2013. Mapping the Social Geographies of Autism–Online and Off-line Narratives of Neuro-Shared and Separate Spaces. *Disability & Society* 28 (3): 367–379.

Billington, Tom. 2016. Critical Autism and Critical Neuroscience: Towards a Science of Research and Practice. In Re-thinking Autism: Diagnosis, Identity and Equality, ed. K. Runswick-Cole, R. Mallett, and S. Timimi. London/Philadelphia: Jessica Kingsley Publishers. pp. 239–251.

Brownlow, Charlotte, and Lindsay O'Dell. 2009. Challenging Understandings of "Theory of Mind": A Brief Report. *Intellectual and Developmental Disabilities* 47 (6): 473–478.

Buber, M. 1937. *I and Thou*. Trans. R.G. Smith. Edinburgh: Clark.

Coeckelbergh, Mark, Cristina Pop, Ramona Simut, Andreea Peca, Sebastian Pintea, Daniel David, and Bram Vanderborght. 2016a. A Survey of Expectations About the Role of Robots in Robot-Assisted Therapy for Children with ASD: Ethical Acceptability, Trust, Sociability, Appearance, and Attachment. *Science and Engineering Ethics* 22 (1): 47–65.

Coeckelbergh, M., C. Pop, R. Simut, A. Peca, S. Pintea, D. David, and B. Vanderborght. 2016b. A Survey of Expectations About the Role of Robots in Robot-Assisted

Therapy for Children with ASD: Ethical Acceptability, Trust, Sociability, Appearance, and Attachment. *Science and Engineering Ethics* 22 (1): 47–65.

Collins, G. 2016. Does a Diagnosis of ASD Help Us to Help a Person with Intellectual Disabilities? In *Re-thinking Autism: Diagnosis, Identity and Equality*, ed. K. Runswick-Cole, R. Mallett, and S. Timimi. London/Philadelphia: Jessica Kingsley Publishers.

Conger, Kate. 2017, June 8. Exclusive: Here's the Full 10-Page Anti-diversity Screed Circulating Internally at Google. *Gizmodo* [Online]

Costescu, Cristina A., and Daniel O. David. 2014. Attitudes Toward Using Social Robots in Psychotherapy. *Erdelyi Pszichologiai Szemle= Transylvanian Journal of Psychology* 15 (1): 3.

Dautenhahn, Kerstin. 2007. Socially Intelligent Robots: Dimensions of Human-Robot Interaction. *Philosophical Transactions: Biological Sciences* 362 (1480): 679–704.

Dautenhahn, Kerstin, and Iain Werry. 2004. Towards Interactive Robots in Autism Therapy: Background, Motivation and Challenges. *Pragmatics & Cognition* 12 (1): 1–35.

Davis, K. 2016. How Rude? Autism as a Study in Ability. In *Re-thinking Autism: Diagnosis, Identity and Equality*, ed. K. Runswick-Cole, R. Mallett, and S. Timimi. London/Philadelphia: Jessica Kingsley Publishers.

Diehl, Joshua J., Lauren M. Schmitt, Michael Villano, and Charles R. Crowell. 2012. The Clinical Use of Robots for Individuals with Autism Spectrum Disorders: A Critical Review. *Research in Autism Spectrum Disorders* 6 (1): 249–262.

Elinor Ochs, Olga Solomon, (2010) Autistic Sociality. Ethos 38 (1):69-92

Esteban, Pablo Gómez, Hoang-Long Cao, Albert De Beir, Greet Van de Perre, Dirk Lefeber, and Bram Vanderborght. 2016. A Multilayer Reactive System for Robots Interacting with Children with Autism. arXiv Preprint arXiv:1606.03875.

Esteban, P.G., P. Baxter, T. Belpaeme, E. Billing, H. Cai, H.L. Cao, M. Coeckelbergh, C. Costescu, D. David, A. De Beir, and Y. Fang. 2017. How to Build a Supervised Autonomous System for Robot-Enhanced Therapy for Children with Autism Spectrum Disorder. *Paladyn, Journal of Behavioral Robotics* 8 (1): 18–38.

Evans, B. 2013. How Autism Became Autism: The Radical Transformation of a Central Concept of Child Development in Britain. *History of the Human Sciences* 26 (3): 3–31.

Goodley, D. 2016. Autism and the Human. In *Re-thinking Autism: Diagnosis, Identity and Equality*, ed. K. Runswick-Cole, R. Mallett, and S. Timimi, 146–158. London/Philadelphia: Jessica Kingsley Publishers.

Grinker, Roy Richard. 2008. *Unstrange Minds: Remapping the World of Autism*. New York: Basic Books.

———. 2010. Commentary: On Being Autistic, and Social. *Ethos* 38 (1): 172–178.

Hacking, I. 2009. Autistic Autobiography. *Philosophical Transactions of the Royal Society of London: Series B, Biological Sciences* 364 (1522): 1467–1473.

Haraway, Donna Jeanne. 2003. *The Companion Species Manifesto: Dogs, People, and Significant Otherness*. Vol. 1. Chicago: Prickly Paradigm Press.

Hashim, R., and H. Yussof. 2017. Humanizing Humanoids Towards Social Inclusiveness for Children with Autism. *Procedia Computer Science* 105: 359–364.

Hassall, R. 2016. Does Everyone with Autism Have the Same Underlying Condition? In *Re-thinking Autism: Diagnosis, Identity and Equality*, ed. K. Runswick-Cole, R. Mallett, and S. Timimi. London/Philadelphia: Jessica Kingsley Publishers.

Haydon-Laurelut, M. 2016. Critical Systemic Therapy: Autism Stories and Disabled People with Learning Difficulties. In *Re-thinking Autism: Diagnosis, Identity and Equality*, ed. K. Runswick-Cole, R. Mallett, and S. Timimi. London/Philadelphia: Jessica Kingsley Publishers.

Hodge, N. 2016. Schools Without Labels. In *Re-thinking Autism: Diagnosis, Identity and Equality*, ed. K. Runswick-Cole, R. Mallett, and S. Timimi. London/Philadelphia: Jessica Kingsley Publishers.

Horlin, Chiara, Marita Falkmer, Richard Parsons, Matthew A. Albrecht, and Torbjorn Falkmer. 2014. The Cost of Autism Spectrum Disorders. *PLoS One* 9 (9): e106552.

Huijnen, Claire A.G.J., Monique A.S. Lexis, Rianne Jansens, and Luc P. Witte. 2016. Mapping Robots to Therapy and Educational Objectives for Children with Autism Spectrum Disorder. *Journal of Autism and Developmental Disorders* 46 (6): 2100–2114.

Kozima, H., C. Nakagawa, and Y. Yasuda. 2005, August. Interactive Robots for Communication-Care: A Case-Study in Autism Therapy. In *Robot and Human Interactive Communication*. ROMAN 2005. IEEE International Workshop on (pp. 341–346). IEEE.

Lasch, Christopher. 1991. *The Culture of Narcissism: American Life in an Age of Diminishing Expectations*. New ed. London/New York: W.W. Norton.

Latif, S. 2016. The Ethics and Consequences of Making Autism Spectrum Disorder. In *Re-thinking Autism: Diagnosis, Identity and Equality*, ed. K. Runswick-Cole, R. Mallett, and S. Timimi. London/Philadelphia: Jessica Kingsley Publishers. pp. 288–299.

Laurelut, Mark Haydon, Saqib Latif, Tom Billington, Gail Simon, Richard Hassall, Brian McCabe, Anne McGuire, Dan Goodley, Rafaela Zorzanelli, and Nick Hodge. 2016. *Re-thinking Autism: Diagnosis, Identity and Equality*. London/Philadelphia: Jessica Kingsley Publishers.

Mallet, R., and K. Runswick-Cole. 2016. The Commodification of Autism: What's at Stake? In *Re-thinking Autism: Diagnosis, Identity and Equality*, ed. K. Runswick-Cole, R. Mallett, and S. Timimi. London/Philadelphia: Jessica Kingsley Publishers.

McCarthy, Anjanie, Kang Lee, Shoji Itakura, and Darwin W. Muir. 2006. Cultural Display Rules Drive Eye Gaze During Thinking. *Journal of Cross-Cultural Psychology* 37 (6): 717–722.

McGuire, A. 2016. Life Without Autism: A Cultural Logic of Violence. In *Re-thinking Autism: Diagnosis, Identity and Equality*, ed. K. Runswick-Cole, R. Mallett, and S. Timimi. London/Philadelphia: Jessica Kingsley Publishers.

Milton, D.E. 2014. Autistic Expertise: A Critical Reflection on the Production of Knowledge in Autism Studies. *Autism*, 18 (7): 794–802.

Muskett, T. 2016. Examining Language and Communication in Autism Spectrum Disorder – In Context. In *Re-thinking Autism: Diagnosis, Identity and Equality*, ed. K. Runswick-Cole, R. Mallett, and S. Timimi. London/Philadelphia: Jessica Kingsley Publishers.

Nadel, Jacqueline, Arnaud Revel, Pierre Andry, and Philippe Gaussier. 2004. Toward Communication: First Imitations in Infants, Low-Functioning Children with Autism and Robots. *Interaction Studies* 5 (1): 45–74.

Ortega, Rafaela, and Clarence Rios. 2016. The Biopolitics of Autism in Brazil. In *Re-thinking Autism: Diagnosis, Identity and Equality*, ed. K. Runswick-Cole, R. Mallett, and S. Timimi, 67–89. London/Philadelphia: Jessica Kingsley Publishers.

Oxford English Dictionary (OED). 1995. The Concise Oxford Dictionary of Current English. First edited by H.W. Fowler and F.G. Fowler. Ninth Edition. Edited by Della Thompson. Oxford: Clarendon Press.

Peca, Andreea, Ramona Simut, Sebastian Pintea, Cristina Costescu, and Bram Vanderborght. 2014. How Do Typically Developing Children and Children with Autism Perceive Different Social Robots? *Computers in Human Behavior* 41: 268–277.

Pop, Cristina Anamaria, Ramona Simut, Sebastian Pintea, Jelle Saldien, Alina Rusu, Daniel David, Johan Vanderfaeillie, Dirk Lefeber, and Bram Vanderborght. 2013. Can the Social Robot Probo Help Children with Autism to Identify Situation-Based Emotions? A Series of Single Case Experiments. *International Journal of Humanoid Robotics* 10 (03): 1350025.

Robins, Ben, R. Te Boekhorst Kerstin Dautenhahn, and Aude Billard. 2005. Robotic Assistants in Therapy and Education of Children with Autism: Can a Small Humanoid Robot Help Encourage Social Interaction Skills? *Universal Access in the Information Society* 4 (2): 105–120.

Rogers, Carl R. 1997 [1967]. *On Becoming a Person: A Therapist's View of Psychotherapy*. London: Constable.

Rosenthal, Robert, and Lenore Jacobson. 1966. Teachers' Expectancies: Determinants of Pupils' IQ Gains. *Psychological Reports* 19: 115–118.

Runswick-Cole, K. 2016. Understanding This Thing Called Autism. In *Re-thinking Autism: Diagnosis, Identity and Equality*, ed. K. Runswick-Cole, R. Mallett, and S. Timimi, 19–29. London/Philadelphia: Jessica Kingsley Publishers.

Runswick-Cole, K., R. Mallett, and S. Timimi. 2016. Future Directions. In *Re-thinking Autism: Diagnosis, Identity and Equality*, ed. K. Runswick-Cole, R. Mallett, and S. Timimi. London/Philadelphia: Jessica Kingsley Publishers.

Russel, G. 2016. Early diagnosis of autism: Is earlier always better? In Re-thinking Autism: Diagnosis, Identity and Equality, ed. K. Runswick-Cole, R. Mallett, and S. Timimi. London/Philadelphia: Jessica Kingsley Publishers. pp. 252–268.

Scassellati, Brian. 2007. How Social Robots Will Help Us to Diagnose, Treat, and Understand Autism. *Robotics Research* 28: 552–563.

———. Implementing Models of Autism with a Humanoid Robot.

Scassellati, Brian, Henny Admoni, and Maja Mataric. 2012. Robots for Use in Autism Research. *Annual Review of Biomedical Engineering* 14: 275–294.

Simon, G. 2016. Thinking Systems: 'Mind' as Relational Activity. In *Re-thinking Autism: Diagnosis, Identity and Equality*, ed. K. Runswick-Cole, R. Mallett, and S. Timimi. London/Philadelphia: Jessica Kingsley Publishers.

Simut, R., Vanderfaeillie, J., Vanderborght, B., Pop, C., Pintea, S., Rusu, A., David, D. and Saldien, J., 2012, March. Is the social robot Probo an added value for Social Story Intervention for children with ASD?. In Human-Robot Interaction (HRI), 2012 7th ACM/IEEE International Conference on (pp. 235-236). IEEE.

Solomon, Olgo. 2012. Doing, Being and Becoming: The Sociality of Children with Autism in Activities with Therapy Dogs and Other People. *The Cambridge Journal of Anthropology* 30 (1): 109–126.

Stawarska, Beata. 2009. *Between You and I: Dialogical Phenomenology*. Athens: Ohio University Press.

Tapus, Adriana, Andreea Peca, Amir Aly, Cristina Pop, Lavinia Jisa, Sebastian Pintea, Alina S. Rusu, and Daniel O. David. 2012. Children with Autism Social Engagement in Interaction with Nao, an Imitative Robot – A Series of Single Case Experiments. *Interaction Studies* 13 (3): 315–347.

Taussig, Michael T. 1993. *Mimesis and Alterity: A Particular History of the Senses*. New York and London: Routledge.

Trimingham, Melissa. 2010. Objects in Transition: The Puppet and the Autistic Child. *Journal of Applied Arts & Health* 1 (3): 251–265.

Timimi, Sami, Neil Gardner and Brian McCabe. 2010. The myth of autism: medicalising men's social and emotional competence. Basingstoke: Palgrave Macmillan.

Timimi, S., and B. McCabe. 2016a. What Have We Learned About the Science of Autism? In *Re-thinking Autism: Diagnosis, Identity and Equality*, ed. K. Runswick-Cole, R. Mallett, and S. Timimi, 30–48. London/Philadelphia: Jessica Kingsley Publishers.

———. 2016b. Autism Screening and Diagnostic Tools. In *Re-thinking Autism: Diagnosis, Identity and Equality*, ed. K. Runswick-Cole, R. Mallett, and S. Timimi, 159–183. London/Philadelphia: Jessica Kingsley Publishers.

CHAPTER 8

Conclusion: Terminating the Machine

8.1 The Social and Cultural Studies of Robots

In this book I have tried to show the ways in which robots are imagined to help children with autism develop social behaviours. I have looked at the ways in which the lifeworlds of children who are described as 'autistic' are compared to asocial machines, and how a back and forth between biomedical psychological sciences of autism and robots confirms this perspective. I first experienced these analogies while I was conducting my doctoral fieldwork in robotics labs at the Massachusetts Institute of Technology in the early 2000s. Elsewhere (Richardson 2015) I have looked at the way in which the boundaries between humans and machines are reconstituted through new kinds of practices in robotics and also in anthropology. I have challenged anthropological frameworks that equate humans with machines or persons with things (Latour 2012; Haraway 1991). I take issue with these approaches by exploring the importance of attachment as crucial for the survival, *and happiness*, of human beings as a species. Rather than find significant intimate relationships are replaceable with machines we find that as humans we are in the best position to support and help other humans develop socially, to be kept alive, and to feel loved and supported. While the therapeutic relationship is different from a friendship or a sexual relationship, it does rely on trust, compassion and empathy between therapist and person. The work of Critical Autism Studies and many other progressive movements that humanise people with differences

K. Richardson, *Challenging Sociality*, Social and Cultural Studies of Robots and AI, https://doi.org/10.1007/978-3-319-74754-5_8

and disabilities allow them to be seen first and foremost as a human being, as a person. Recently, while speaking to a clinical psychologist I was told in no uncertain terms that autism was a 'mental disorder'. He was correct, if you look at the most influential tool for psychologists and psychiatrists—it is called The Diagnostic Manual for Mental Disorders (DSM) and autism is in there, along with hundreds of other conditions. New 'mental disorders' are emerging every year. In DSM-5 (American Psychiatric Association 2013) new 'mental disorders' include 'Premenstrual Dysphoric Disorder' (menstrual cramps) and 'Hoarding Disorder' (hoarding), and 'Caffeine Withdrawal' as disorders that produce mental distress. It is the biomedical and behavioural models of autism that are made use of by robotics scientists to develop robots for children with autism, by making claims that robots are more accessible to children with autism than are other human beings. Perhaps it is because the models for making sense of autistic persons in biopsychiatry and psychology are so mechanical and reductivist to begin with, that it is an easy journey from the psychological clinics to the roboticists labs. If this is so, why? Is it because there is a similar egocentric and mechanical idea that underlies autism and robots? In this book I have tried to raise these questions and problematize this journey from child with autism to robot.

In Chap. 1 I described the rise of robots to be socially assistive to children with autism. The problem is not so much the development of robots as tools. Robots are types of machines and machines are complex tools. Human beings make and use tools in order to enhance existence and transform it, but tools can be produced that are harmful, such as nuclear weapons or maybe even robots that are made to interact with children with autism. As this technology is at its earliest stages, we do not know the long-term effects of regularly exposing children to interactaction with robots and encouraging them to socially engage with them. We know that television and computer games, mobile phones and personal computers have had a tremendous effect on the way children socially interact, sometimes for the worse and the better (Kizel 2017; Wegerif and Major 2018; Trausan-Matu 2017).

The problem I address in this book is the surreptitious associations between autistic people and machines, as though autism is analogical to a machine*like* state. This belief that autistic people shun human association in favour of things was an idea presented in Leo Kanner's original paper in 1943. Since then this idea has been picked up and made use of in a variety of different ways from authors as diverse as Uta Frith, Simon Baron-Cohen, Michael Rutter and Peter Hobson. It represents an underlying idea about people with autism that fed directly into the making of humanlike robots as therapeutic tools for children with autism to help these children develop social skills.

However, what autism is, or who has it, or how many people have it, has changed considerably since the term autism was first coined by Eugen Bleuler in 1913 to describe schizophrenics as showing autistic behaviour. It was Kanner that transformed autism and created a separate categorisation of autism. Autism has undergone various shifts over time. One significant shift was the expansion of the category of autism in the DSM-IV (1951). In the 1990s the category of Asperger's Syndrome was added, along with a wider categorisation of autism, creating what some historians of medicine point to an explosion in autism diagnosis. The category of autism changed again with the publication of the DSM-5, and this time Asperger's was removed from the manual. If autism cannot be fixed in time and space by the psychological sciences, it cannot be a fixed target for roboticists either. Much of the basis for arguments for robots is built on biomedical understandings of autism that have come under considerable critical scrutiny. Moreover, the basis for these claims from roboticists comes from small scale experimental studies, that have relied on coding the behaviours of children with autism interacting with the robot. These studies rely on observational material when the child is present with a robot, and none have proven that any skills are meaningfully transferred from the interaction with a robot into the social life of the child in other settings.

Chap. 3 is a review of a pilot study exploring the effectiveness of using a humanoid robot to help children with autism develop social behaviours. The robot was provided by a British university lab and the experimental design by an autism expert. The formal part of the experiment ran over two months in 2010 in London at a specialist school for children with autism. Parents were invited to participate in the experiment via a school letter, and 29 children participated in the study. The experiment ran in three sections. The experimental group interacted with a humanoid robot for up to five minutes over nine sessions. The control group interacted with a brightly coloured toy crane for up to five minutes over nine sessions. The experiment was punctuated at the beginning and at the end with a play protocol to establish the child's social awareness and behaviours.

In Time 1, the children were video-taped playing with toys and responding to three tasks which included; responding to the blowing of bubbles; 2. responding to the blowing of a balloon; and 3. responding to hiding of an object. In all these three tasks the children were measured on their engagement. This Time 1 play protocol was performed for each child at the beginning and at the end of the experiment. The experimental (humanoid robot) and control (crane) group participants were randomly assigned.

At the end of the experiment, the same play protocol was reproduced and coded for any difference in behaviour. The researchers found no difference in behaviour. The results of the experiment were never published. It was during this post-experiment period that I learned about the way scientists shelve negative results. The lack of findings was put down to a variety of reasons, including lack of standardisation of the environment; lack of control over environmental noises; and the shortness of the sessions (only nine sessions over a period of two months for up to five minutes for each interaction).

As an anthropologist and ethicist of the ethics and culture of robotics, I describe the experiment, and the lack of results. I wonder sometimes what would have happened had the results shown a positive correlation between the robot use and desired behaviours in the children with autism. I evaluate this experiment in light of the millions of euros that are now put into developing robots for children with autism.

In Chap. 4 I take up the theme of autism as a withdrawn state, and explore this through the psychological and psychoanalytic use of the mythic tale of Echo and Narcissus in Ovid's Metamorphoses, to look at the underlying problem in Western society of an egocentric focus which reduces the you to an It. The primary focus of Western individualism on the ego, and the development of knowledge that is driven by egocentric individualism, produce limited ways of knowing about otherness, because the otherness is always reduced to a thing. What began as an analogy and metaphor: the machine turns into a powerful othering practice. Children with autism are not outside human interpersonal relationships, but part of them.

In Chap. 5 I wanted to look at attachment theories and the history of parenting and mother-child attachment. Autism and attachment have been unhappy associates, as mothers were blamed for their children's autism. Mothers have been blamed for turning infants and children into psychotic and neurotic personalities. Mother-love and mother-hate run in parallel. The value of attachment theory is the insights it can give us into the life of children, which have been largely ignored in modern cultures or children dehumanised as other—sinful, wilful, sexually perverse.

Children for much of history have experienced severe physical punishment at the hands of parents and other adults (Miller 1987; Shorter 1975; Stone 1979). Attachment theory began to open up a compassionate perspective towards children and look at the way human subjectivity was informed by parental influences. Unfortunately, many of the first attachment studies took as given particular kinds of normative arrangements, such the role of mother as primary caregiver. Mothers play an extremely important role in the lived lives of human beings, but this role can be supported by including fathers as part of caregiving. For a child, the value of a secure and enduring relationship, free from violence and abuse, enhanced by a culture that supports education and enrichment of the child, and fostered by parents' participation in their child's life as supported by socially progressive governmental policies cannot be overstated. In a culture that is dominated by excessive materialism and consumerism, relationships between people are devalued.

Robots for children with autism are presented as therapeutic tools, for now. There are also many roboticists claiming that robots could replace human relationships such as a friend, or even an intimate partner. How long before roboticists promote the use of robots as friends to children? Robots as significant others represents a new stage in the commodification of human relationships. Advertising promotes products by making claims about what a product can do to enhance the status of a person. Robots are commercially produced goods—every part of the robot is assembled with a manufactured commodity. Humans by contrast are not completely manufactured and commercial. Marketing and advertising companies encourage consumers to develop relationships with products, even to 'bond' with the product. In some ways, the robot seems outside of typical commercialism, but we should restore it back into commercial culture. If robots are promoted as viable substitutes to significant intimate relationships or as alternatives to human therapists, they should be promoted ethically and accurately to the public and consumers. However, attachment studies affirm the importance of interpersonal species-specific relationships. Harry Harlow's studies tell us absolutely that 'surrogate' machines cannot produce well-adjusted monkeys.

Attachment has a meaning in general language that is different from its specific meaning in psychology. Attachment is how many robotics scientists describe the relationships between people and robots. Attachment is also a term used by marketing and advertising agencies to increase consumer feelings towards certain products. Can attachment really refer to this interpersonal experience between parents and infants and simultaneously be the same term appropriate to consumers' use of mobile phones or robots? This transference of attachment theory out of the domain exploring interpersonal relationships can now be applied to the way human beings relate to commercially produced products. Humanoid robots are commercially produced products. They are bought in parts and put together in robotics labs, or as standalone commercial platforms that can be modified by researchers and innovators. The ease with which robotics researchers and marketing agencies speak about 'attachment' to their products should raise concerns. The first concern is an appropriation and misuse of the original attachment theory by the fields of marketing and advertising to sell more products. Attachment studies explore infant and child development and how the appropriate attachment form supports growth and maturation. Attachment to commercially produced mobile phones or robots does not support development, growth or maturation in this way. Without mothers, fathers and other significant caregivers in a child's life children would not exist. Caring for children is not merely a mechanical exchange of services, and these 'services' cannot be mapped onto a machine.

Finally, human-human interaction can never be mapped onto human-robot interaction because they are ontologically divergent. Machines are artefacts and have no ontological experience. Human beings project onto artefacts their fantasies, thoughts and feelings. Good robotic puppeteers can even engineer machines to mimic behaviours. Egocentric relationships are writ large as the norm—what happens in the mind of the individual, what they are thinking and feeling, is what is important, I am frequently told by roboticists. If they believe the robot is alive, and their friend, then it is alive. This underlying absence of the other, of the you as a thou, and not as an it—a thing to be used drives the model of relationship in robotics.

Should an infant not have a human caregiver, their chances of survival are almost non-existent, and certain kinds of animals such as dogs or wolves would be better caregivers than machines, as while each species has its own unique form of attachment, all living beings have a species-specific sociality. As far as we know, no new born infant has ever been raised by any other species, and if a young infant was to fall into the hands of an animal caregiving family, if the child survived, it would develop along the lines of the species in which it was developing.

Some people propose that robots will be able to do the caretaking roles of parents, and infants of the future, some speculate, will spend long periods of time in the presence of a machine.

In Chap. 7 we take a look at new and exciting research to come out the Critical Autism Studies school as a collective of psychiatrists, autistic adults, parents of children and adults with autism, sociologists, anthropologists and historians of autism (Runswick-Cole et al. 2016; Timimi and McCabe 2016a, b; Russell 2016; Billington 2016; Haydon-Laurelut 2016; Ortega and Rios 2016; Muskett 2016) and others (Ochs and Solomon 2010; Grinker 2010; O'Dell et al. 2016; Bagatell 2010; Evans 2013; Milton 2014; Robinson 2014; 2015) who are problematizing mainstream orthodoxy. These scholars and activists are asking important questions; they are David compared to the Goliath that is the powerful biosciences industries. Autism is a multibillion-dollar industry which pushes an agenda towards bioscience and away from humanistic not-for-profit paradigms. An important dialogue needs to take place between these different communities, but the Critical Autism Studies perspectives are marginalised.

Critical Autism Studies authors, anthropologists, sociologists, autism activists, parents present new radical perspectives inviting us to *re-think autism*, from its diagnosis, to its clinical specification, and industry and commercialisation of autism.

8.2 I and Thou

The use of Martin Buber's model of I and Thou dialogue that forms the ethical foundation of this book is built around an ethics of being with others, rather than using others, or 'itifying' others by turning them into objects.

I wanted to get to a place in this book to ask, what might it be like to see people who may have autistic traits without pathologising and othering people with autism? What might it be like to look at the potential of a technological tool like a robot to be useful among people with autistic like traits without creating analogies between people with autism and machines as sharing similar kinds of states?

As a tool, robots have the potential to engage, interest and contribute to the education of children of all kinds of abilities.

States can make these kinds of decisions in how they decide to invest their resources. Also, the Critical Autism Studies research groups have opened up important avenues of research and have raised crucial humanistic questions about ensuring that the humanity of people with autism is never lost. Machines are good models for thinking about tools and appliances, but not for thinking about human beings. It is time to take the machine out of the human being and terminate the egocentric tradition in favour of I-you interrelatedness.

I-you interrelatedness refers to what is happening within and between human beings. There is no I in the absence of a you. There is no egocentric universe in which Man is alone as fictionalised in the individualistic paradigm. Therefore humans do not merely project onto others, but are interpersonal with others. The psychological sciences and biomedical approaches have broken up the interpersonal whole of the human into parts, and removed history, politics and economics from the narratives they produce. The experiment becomes a way in which this 'truth' is produced. The science and technological community do not report all the results from their experiences, but present the findings that are positive, excluding the results of null experiments. Robotics, after all, is about the development of machines, which are reductive and mechanical. This objectification of the 'autistic person' through psychiatry and bio-medical approaches to autism is made explicit by the development of robots for children.

8.3 Robots as Useful Tools Without Pathologising Autism

As Ortega, Zorzanelli and Rios write of autistic narratives, their power and sociological purpose, 'narratives are not simply representations of reality, but serve to structure human experience and to organise our actions in cultural contexts in which they take place' (p. 76). Taking this idea as our starting point, we could ask, is it possible to develop robots for all children diagnosed with autism or developmental and other intellectual disabilities without pathologising them?

Is it possible to develop new technologies without referring to pathological ideas about human beings, and the experience of being human that is classified as 'autistic'? The pervasive characterisation of children with autism, their personalities, behaviours and traits as asocial and machine*like* is a problem. It is what Goodley (2016) referred to as creating an 'Other'; drawing on postcolonial studies he argues:

> what other conception of the human might exist had the white man not got there first! This is not to deny that humanness has a long, complex history of development (encompassing multi-ethnic contributions) but to think how the contemporary human might be explained in relation to a number of genealogical origins. (p. 149)

If we began to see those diagnoses with a label of autism less as a fixed and stable category, but more as a culturally shifting category? One cannot study autism without studying the way it which it the social and cultural norms, values and attitudes around autism, that have been shaped by a powerful biomedical approach which has viewed difference as pathological. Autism studies must take account of the ways in which powerful bodies produce knowledge systems, biomedicines, products, services, biographies and other commercially products outputs from autism narratives?

Taking into account the way in which technology is increasingly a part of many children's lives in European and North American contexts, there needs to be a wider discussion about the benefits and use of personalized communication technologies and the long-term impacts on children's development. As Robinson (2014) explains, examining the premises that go into different kinds of service provision for adults with Asperger's is important. Empowering adults with Asperger's to come up with their own narratives in spaces that are safe and encouraging is profoundly important. Is a narrative that children with autism prefer machines over other human beings a positive supportive narrative for children with autism? These *othering* premises need to be addressed in autism and robot narratives and practices to first identify if, when these underlying ideas of autism as a systemising state are removed, there is any value in using robots for therapeutic purposes at all. This approach would call into question all the biomedical assumptions that drive a multimillion-dollar industry. If the systemising and biomedical notions of autism are removed, do the robots help? Is it possible to look in different ways at how robots could interest the children who use them—not just the adults who create them—without losing sight of the ways in which robots could be helpful as therapeutic, educational or play tools for children.

In the twenty first century, there are more studies, research books, pamphlets, guidelines, experiments and biographies of childhood than ever before in the human lived experience. We know that even if at other times children were not seen as human, when they were beaten, or abused by adults, these actions harm children. We also know that children with developmental disabilities need different kinds of support and help to flourish into healthy and fulfilled adults. Attachment studies reveal the importance of a loving origin story, and the need to create a safe world for children so they are not harmed by adults. Moreover, the ethics discussed here, of I-you interrelatedness and humanities, needs to inform the approaches we employ towards disabilities, abilities and difference.

Starting from different premises about the value of all human beings, and a critical approach to the othering language of autism or other kinds of disabilities and difference, we can change the conversation about autism in wider society. The Critical Autism Studies network, I believe, has begun this conversation—and it needs to be included into approaches to the ethics of robots for children with autism. What is fundamental to all human beings is the importance of a loving origin story.

Bibliography

American Psychiatric Association. 2013. *Diagnostic and Statistical Manual of Mental Disorders (DSM-5®)*. American Psychiatric Pub.

Bagatell, Nancy. 2010. From Cure to Community: Transforming Notions of Autism. *Ethos* 38 (1): 33–55.

Billington, Tom. 2016. Critical Autism and Critical Neuroscience: Towards a Science of Research and Practice. In *Re-thinking Autism: Diagnosis, Identity and Equality*, ed. K. Runswick-Cole, R. Mallett, and S. Timimi, 239–251. London/Philadelphia: Jessica Kingsley Publishers.

Bleuler, E. 1951. Autistic Thinking. In *Organization and Pathology of Thought: Selected Sources*, ed. D. Rapaport, 399–437. New York: Columbia University Press.

Davies, Kim. 2016. How Rude? Autism as a Study in Ability. In *Re-thinking Autism: Diagnosis, Identity and Equality*, ed. K. Runswick-Cole, R. Mallett, and S. Timimi, 132–145. London/Philadelphia: Jessica Kingsley Publishers.

Evans, Bonnie. 2013. How Autism Became Autism: The Radical Transformation of a Central Concept of Child Development in Britain. *History of the Human Sciences* 26 (3): 3–31.

Goodley, D. 2016. Autism and the Human. In *Re-thinking Autism: Diagnosis, Identity and Equality*, ed. K. Runswick-Cole, R. Mallett, and S. Timimi, 146–158. London/Philadelphia: Jessica Kingsley Publishers.

Grinker, Roy Richard. 2010. Commentary: On Being Autistic, and Social. *Ethos* 38 (1): 172–178.

Haraway, Donna Jeanne. 1991. *Simians, Cyborgs and Women: The Reinvention of Nature*. London: Free Association Books.

Haydon-Laurelut, M. 2016. Critical Systemic Therapy: Autism Stories and Disabled People with Learning Difficulties. In *Re-thinking Autism: Diagnosis, Identity and Equality*, ed. K. Runswick-Cole, R. Mallett, and S. Timimi, 221–238. London/Philadelphia: Jessica Kingsley Publishers.

Kizel, A. 2017. I–Thou Dialogical Encounters in Adolescents' WhatsApp Virtual Communities. *AI & SOCIETY*: 1–9.

Latour, Bruno. 2012. Gabriel Tarde and the End of the Social. In *The Social in Question. New Bearings in History and the Social Sciences*, ed. Patrick Joyce, 117–132. London: Routledge.

Miller, Alice. 1987. *The Drama of Being a Child: And the Search for the True Self*. London: Virago.

Milton, D.E. 2014. Autistic Expertise: A Critical Reflection on the Production of Knowledge in Autism Studies. *Autism: The International Journal of Research and Practice* 18 (7): 794–802.

Muskett, Tom. 2016. Examining Language and Communication in Autism Spectrum Disorder – In Context. In *Rethinking Autism: Diagnosis, Identity and Equality*, ed. K. Runswick-Cole, R. Mallett, and S. Timimi, 300–316. London/Philadelphia: Jessica Kingsley Publishers.

O'Dell, Lindsay, Hanna Bertilsdotter Rosqvist, Francisco Ortega, Charlotte Brownlow, and Michael Orsini. 2016. Critical Autism Studies: Exploring Epistemic Dialogues and Intersections, Challenging Dominant Understandings of Autism. *Disability & Society* 31 (2): 166–179.

Ochs, Elinor, and Olga Solomon. 2010. Autistic Sociality. *Ethos* 38 (1): 69–92. https://doi.org/10.1111/j.1548-1352.2009.01082.x.

Ortega, Rafaela, and Clarence Rios. 2016. The Biopolitics of Autism in Brazil. In *Re-thinking Autism: Diagnosis, Identity and Equality*, ed. K. Runswick-Cole, R. Mallett, and S. Timimi, 67–89. London/Philadelphia: Jessica Kingsley Publishers.

Richardson, Kathleen. 2015. *An Anthropology of Robots and AI: Annihilation Anxiety and Machines*. Vol. 20. New York: Routledge.

Robinson, J. 2014. Participatory Research with Adults with Asperger's Syndrome: Using Spatial Analysis to Explore How They Make Sense of Their Experience. Doctoral Dissertation, De Montfort University, Leicester.

———. 2014; 2015. *Participatory Research with Adults with Asperger's Syndrome: Using Spatial Analysis to Explore How They Make Sense of Their Experience*. De Montfort University.

Runswick-Cole, K., R. Mallett, and S. Timimi. 2016. Introduction. In *Re-thinking Autism: Diagnosis, Identity and Equality*, ed. K. Runswick-Cole, R. Mallett, and S. Timimi, 110–132. London/Philadelphia: Jessica Kingsley Publishers.

Russel, G. 2016. Early Diagnosis of Autism: Is Earlier Always Better? In *Re-thinking Autism: Diagnosis, Identity and Equality*, ed. K. Runswick-Cole, R. Mallett, and S. Timimi, 252–268. London/Philadelphia: Jessica Kingsley Publishers.

Shorter, E. 1975. *The Making of the Modern Family*. New York: Basic Books.

Stone, L. 1979. *The Family, Sex and Marriage in England 1500–1800*. Vol. 43. Harmondsworth: Penguin.

Timimi, S., and B. McCabe. 2016a. What Have We Learned About the Science of Autism? In *Re-thinking Autism: Diagnosis, Identity and Equality*, ed. K. Runswick-Cole, R. Mallett, and S. Timimi, 30–48. London/Philadelphia: Jessica Kingsley Publishers.

Timimi, S., and B. McCabe. 2016b. Autism Screening and Diagnostic Tools. In *Re-thinking Autism: Diagnosis, Identity and Equality*, ed. K. Runswick-Cole, R. Mallett, and S. Timimi, 159–183. London/Philadelphia: Jessica Kingsley Publishers.

Trausan-Matu, S. 2017. Is It Possible to Grow an I–Thou Relation with an Artificial Agent? A Dialogistic Perspective. *AI & SOCIETY*: 1–9.

Wegerif, R., and L. Major. 2018. Buber, Educational Technology, and the Expansion of Dialogic Space. *AI & SOCIETY*: 1–11.

Index[1]

[1] Note: Page numbers followed by 'n' refer to notes.

K. Richardson, *Challenging Sociality*, Social and Cultural Studies of Robots and AI, https://doi.org/10.1007/978-3-319-74754-5

The manufacturer's authorised representative in the EU is Springer Nature Customer Service Centre GmbH, Europaplatz 3, 69115 Heidelberg, Germany. If you have any concerns regarding our products, please contact ProductSafety@springernature.com

Printed and bound by CPI Group (UK) Ltd, Croydon, CR0 4YY

22/09/2026

02220950-0009